Physics and Technology of Semiconductor Devices

A. S. GROVE

Intel Corporation, Mountain View

University of California, Berkeley

John Wiley and Sons, Inc., New York · London · Sydney

Copyright © 1967 by John Wiley & Sons, Inc.

All rights reserved. This book or any part thereof must not be reproduced in any form without the written permission of the publisher.

Library of Congress Catalog Card Number: 67-17340
Printed in the United States of America

ISBN 0 471 32998 3

14 15 16 17 18 19 20

IMPORTANT ELECTRONIC PROPERTIES OF SEMICONDUCTORS

(For other properties, see Table 4.1 on page 102.)

	Ge	Si	GaAs
Energy gap (ev)	0.67	1.1	1.4
Effective density of states (cm^{-3}) conduction band N_c valence band N_v	1.04 × 10^{19} 6.0 × 10^{18}	2.8 × 10^{19} 1.04 × 10^{19}	4.7 × 10^{17} 7.0 × 10^{18}
Intrinsic carrier concentration n_i (cm^{-3})	2.4 × 10^{13}	1.45 × 10^{10}	~9 × 10^{6}
Lattice (intrinsic) mobilities (cm^2/vsec) electrons holes	3900 1900	1350 480	8600 250
Dielectric constant	16.3	11.7	12
Breakdown field (v/μ)	~8	~30	~35

Physics and Technology of Semiconductor Devices

Preface

The 1950's was the decade in which semiconductor devices rose to prominence and attained great industrial significance. In that decade, most of the research and engineering work was directed to germanium. The 1960's can be considered as the decade in which silicon semiconductor devices and integrated circuits made by the *planar technology* overtook germanium devices.

The purpose of this book is to provide an introduction to the physics and technology of *planar silicon devices*, i.e., devices made by the planar technology. To be sure, the physical principles underlying the fabrication and the operation of these devices do not differ from those underlying the fabrication and the operation of devices made from other semiconductors by other technologies. However, problems viewed in general terms tend to be very difficult to solve. To render a problem tractable, we must concentrate on its most important features. What is important for an alloy germanium diode, for instance, may not be important for a planar silicon diode. As a result, approximations which describe well the characteristics of the former often give results in very poor agreement with observations on the latter. Most if not all books in the past have tacitly concentrated on those factors that are important in relation to germanium alloy devices. In this one, the emphasis is on those which are important in dealing with planar silicon devices.

After a brief description of the planar technology, the text is divided into three parts. The first one deals with those processes of *solid-state technology* which are most intimately involved in determining the electrical characteristics of semiconductor devices. The three processes discussed—vapor-phase growth, thermal oxidation, and solid-state diffusion—are the means by which the semiconductor crystal is turned into a device of desired characteristics.

The second part deals with *semiconductors and semiconductor devices*. A summary of the most important results of the band theory of solids is followed by a discussion of semiconductors under non-equilibrium conditions. Then *p-n* junctions, which are the building block of most important semiconductor devices, are considered in detail followed by a treatment of junction transistors and junction field-effect transistors.

One of the principal advantages of the planar technology is that it results in excellent control of the semiconductor surface. This control has made the surface a designable part of semiconductor devices and has also led to the rapid development of new types of devices. One of these, the surface field-effect transistor, is already second in importance only to the junction transistor among active devices. Thus the third part of this text considers *surface effects and surface-controlled devices*. This part includes discussion of the theory of semiconductor surfaces, surface effects on *p-n* junctions, the surface field-effect transistor, and the status of the silicon-silicon dioxide system.

The text is principally intended for use by senior undergraduate or first-year graduate students in electrical engineering and in materials science. For this reason I have attempted to keep the most important physical principles always in the forefront and to use simple physical models wherever possible. However, I have also attempted to make the book remain useful to the student when he becomes a practicing engineer or scientist. To serve this purpose, *principles* are generally illustrated with actual numerical computations or with experimental measurements. Thus, in addition to boosting the readers' confidence in these principles, the illustrations also contain *numbers* which, at a later time, can be used in design calculations.

The dual nature of this book was also helped by the way it developed. The material contained in it is based on a series of lectures given at Fairchild Semiconductor to scientists and engineers who were engaged in research, development, manufacturing, and application work involving semiconductor devices and integrated circuits, and on a graduate course I taught at the University of California, Berkeley. The material, in manuscript form, was then used in the Fairchild course in the following year, as well as by Professors C. A. Mead, California Institute of Technology, and P. O. Lauritzen, University of Washington, in their undergraduate electrical engineering courses.

Preface

In the writing of this book I benefited very significantly from the many suggestions of the reviewers: D. J. Fitzgerald, E. H. Snow, L. Vadasz, and R. J. Whittier of Fairchild Semiconductor; and Professors P. O. Lauritzen, University of Washington, C. A. Mead, California Institute of Technology, J. L. Moll, Stanford University, and W. G. Oldham, University of California, Berkeley. I am further indebted to J. P. Bissell for his assistance in evaluating device characteristics, and to Miss S. J. Steele and Mrs. Dorothy Unruh for their help with the calculations and illustrations. Finally, I would like to express my appreciation to the management of Fairchild Semiconductor for providing an environment in which the writing of this book was possible.

A. S. GROVE

Palo Alto, California
February 1967

Contents

Introduction: The Planar Technology 1

I. SOLID-STATE TECHNOLOGY

Chapter 1 Vapor-Phase Growth 7

    1.1 Kinetics of Growth 10
    1.2 Gas-Phase Mass-Transfer 13
    1.3 Some Properties of Gases 18

Chapter 2 Thermal Oxidation 22

    2.1 Kinetics of Oxide Growth 23
    2.2 Space-Charge Effects on Oxidation 31

Chapter 3 Solid-State Diffusion 35

    3.1 Flux 36
    3.2 The Transport Equation 41
    3.3 Diffused Layers 43
    3.4 Deviations from Simple Diffusion Theory 58
    3.5 Redistribution of Impurities during Thermal Oxidation . 69
    3.6 Diffusion through a Silicon Dioxide Layer (Oxide Masking) 75
    3.7 The Redistribution of Impurities in Epitaxial Growth . 78

II. SEMICONDUCTORS AND SEMICONDUCTOR DEVICES

Chapter 4 Elements of Semiconductor Physics 91

    4.1 The Band Theory of Solids 91
    4.2 Electrons and Holes in Semiconductors 95
    4.3 Fermi-Dirac Distribution Function 98
    4.4 Important Formulas for Semiconductors in Equilibrium 100
    4.5 Transport of Electrons and Holes 106

Chapter 5 Semiconductors under Non-Equilibrium Conditions 117

    5.1 Injection 117
    5.2 Kinetics of the Recombination Process 127
    5.3 Lifetime in Low-Level Injection 134
    5.4 Surface Recombination 136
    5.5 Origin of Recombination-Generation Centers 140

Chapter 6 $p\text{-}n$ Junctions 149

    6.1 Electrostatics 152
    6.2 Space-Charge Region for Step Junctions 153
    6.3 Space-Charge Region for Linearly Graded Junctions . . 163
    6.4 Space-Charge Region for Diffused Junctions 167
    6.5 Capacitance-Voltage Characteristics 169
    6.6 Current-Voltage Characteristics 172
    6.7 Junction Breakdown 191
    6.8 Transient Behavior 201

Chapter 7 Junction Transistors 208

    7.1 Principles of Transistor Action 209
    7.2 Currents Flowing in a Transistor; Current Gain 214
    7.3 Limitations and Modifications of the Simple Theory . . 222
    7.4 Base Resistance 228
    7.5 Maximum Voltage Limitations 230
    7.6 Minimum Voltage Limitations 234
    7.7 Thermal Limitation 236

Chapter 8   Junction Field-Effect Transistors   243

8.1 Principles of Operation   244
8.2 Characteristics of Junction Field-Effect Transistors . . .   248
8.3 Modifications of the Simple Theory   253

III. SURFACE EFFECTS AND SURFACE-CONTROLLED DEVICES

Chapter 9   Theory of Semiconductor Surfaces   263

9.1 Characteristics of Surface Space-Charge Regions—Equilibrium Case   264
9.2 The Ideal MIS (or MOS) Structure   271
9.3 Effect of Work Function Difference, Charges, and States on MOS Characteristics   278

Chapter 10   Surface Effects on p-n Junctions   289

10.1 Characteristics of Surface Space-Charge Regions—Non-Equilibrium Case   290
10.2 Gate-Controlled Diode Structure   296
10.3 Recombination-Generation Processes in the Surface Space-Charge Region   298
10.4 Field-Induced Junctions and Channel Currents   305
10.5 Surface Effects on Junction Breakdown Voltage   311

Chapter 11   Surface Field-Effect Transistors   317

11.1 Principles of Operation   318
11.2 Characteristics of Surface Field-Effect Transistors   321
11.3 Modification of the Simple Theory   327
11.4 Other Types of Surface Field-Effect Transistors   329

Chapter 12   Properties of the Silicon-Silicon Dioxide System   334

12.1 Fast Surface States   335
12.2 Space Charge within the Oxide   337

12.3 Surface-State Charge	342
12.4 Barrier Energies	345
12.5 Surface Mobility	346
12.6 Conduction on Oxide Surfaces	347
12.7 Other Insulators	350

Index **357**

List of Symbols

a	impurity concentration gradient at the junction, $\left.\dfrac{dC}{dx}\right|_{x=x_j}$
A	coefficient in the general relationship for the thermal oxidation of silicon
A_J	cross-sectional area of metallurgical p-n junction
A_s	depleted surface area
B	coefficient in the general relationship for the thermal oxidation of silicon; also referred to as parabolic rate constant
B	transistor base factor
BV	junction breakdown voltage
BV_{CBO}	collector-base junction breakdown voltage, with emitter open
BV_{CEO}	collector-to-emitter breakdown voltage, with base open
C	concentration
C	capacitance per unit area
C^*	equilibrium concentration of oxidant in oxide film
C_B	bulk concentration
C_f	concentration of external doping impurity at the surface of an epitaxial film
C_G	concentration of a species in the bulk of the gas
C_G	total gate capacitance
C_i	concentration of the oxidant at the oxide-silicon interface
C_o	concentration of the oxidant at the gas-oxide interface
C_o	capacitance per unit area of oxide layer
C_S	surface concentration
C_s	capacitance per unit area of the surface space-charge region
C_{sub}	impurity concentration within the substrate in epitaxial growth

C_T	total concentration of molecules in a gas
C_T	total concentration of ionized impurities in a crystal
d	thickness of metallurgical channel of junction field-effect transistor
D	diffusion coefficient
D_{eff}	effective diffusion coefficient
D_G	gaseous diffusion coefficient
D_n	diffusivity of electrons
D_o	diffusivity of impurities in oxide
D_p	diffusivity of holes
D_{st}	density of uniformly distributed surface recombination-generation centers (per unit area and energy)
D_t	concentration of uniformly distributed bulk recombination-generation centers (per unit volume and energy)
$\mathscr{E}$	electric field
E	electron energy
E	transistor emitter factor
E_a	activation energy
E_c	electron energy at conduction band edge
$\mathscr{E}_{\text{crit}}$	critical electric field at breakdown
E_F	electron energy at the Fermi level
E_{Fn}	electron energy at quasi-Fermi level for electrons
E_{Fp}	electron energy at quasi-Fermi level for holes
E_G	width of the forbidden gap
E_i	electron energy at the intrinsic Fermi level
$\mathscr{E}_s$	electric field at the semiconductor surface
E_t	energy level of recombination-generation center
E_v	electron energy at valence band edge
F	flux
F_n	flux of electrons
F_p	flux of holes
f	probability of occupation by an electron
f_o	maximum frequency of operation
G_L	rate of generation of electron-hole pairs due to absorbed light (per unit time and volume)
G_o	conductance of the metallurgical channel of a junction field-effect transistor
G_{th}	rate of generation of electron-hole pairs in dark (per unit time and volume)

List of Symbols

g	channel conductance
g_m	transconductance
g_{msat}	transconductance in the saturation region
H	Henry's-law constant
h	gas-phase mass-transfer coefficient in terms of concentrations in the solid
h_{FB}	common-base current gain
h_{FE}	common-emitter current gain
h_{fe}	small-signal common-emitter current gain
h_G	gas-phase mass-transfer coefficient in terms of concentrations in the gas
h_{th}	gas-phase heat-transfer coefficient
I	current
I_B	base current
I_C	collector current
I_{CBO}	collector-base reverse current, with emitter open
I_{CEO}	emitter-to-collector reverse current, with base open
I_D	drain current
I_{diff}	diffusion current
I_{Dsat}	drain current in saturation region
I_E	emitter current
I_F	forward current
I_{gen}	generation current
I_p	hole current
I_R	reverse current
I_{rec}	recombination current
k	Boltzmann's constant
k_s	surface reaction rate constant
k_{th}	thermal conductivity
K	dielectric constant
K_o	dielectric constant of oxide
K_s	dielectric constant of semiconductor
L	channel length
L_n	diffusion length of electrons
L_p	diffusion length of holes

m	segregation coefficient of impurity at the oxide-silicon interface
m^*	effective mass
M	multiplication factor
n	electron concentration
n_i	intrinsic carrier concentration
n_n	concentration of electrons in an n-type semiconductor
n_{no}	concentration of electrons in an n-type semiconductor in equilibrium
n_p	concentration of electrons in a p-type semiconductor
n_{po}	concentration of electrons in a p-type semiconductor, in equilibrium
n_s	electron concentration at the surface
N_A	concentration of acceptor impurities
N_c	effective density of states in the conduction band
N_D	concentration of donor impurities
N_{st}	density of surface recombination-generation centers per unit area
N_t	concentration of bulk recombination-generation centers per unit volume
N_v	effective density of states in the valence band
p	pressure
p	hole concentration
p_n	concentration of holes in an n-type semiconductor
p_{no}	concentration of holes in an n-type semiconductor in equilibrium
p_p	concentration of holes in a p-type semiconductor
p_{po}	concentration of holes in a p-type semiconductor, in equilibrium
p_s	hole concentration at the surface
q	magnitude of electronic charge, 1.6×10^{-19} coulomb
Q	total amount of impurities per unit area in the solid
Q_B	total number of impurities per unit area in the base region of a transistor
Q_B	charge per unit area within the surface depletion region at the onset of strong inversion
$Q_{B,o}$	charge per unit area within the surface depletion region at the onset of strong inversion, in equilibrium
Q_G	charge per unit area on the gate
Q_o	charge per unit area within the oxide
Q_n	charge per unit area due to electrons in the inversion layer

List of Symbols

Q_s	charge per unit area in the semiconductor
Q_{ss}	fixed surface-state charge density per unit area
R	ideal gas-law constant
R	total recombination rate (per unit time and volume)
R	transistor recombination factor
R_d	drain series resistance
Re	Reynolds number
R_s	source series resistance
R_{th}	thermal resistance
$R_\square$	sheet resistance of a square
$r_B{'}$	base spreading resistance
r_{SC}	series resistance of the collector region
r_{SE}	series resistance of the emitter region
s	surface recombination velocity
s_{max}	maximum surface recombination velocity
s_o	surface recombination velocity of a surface without a surface space-charge region
T	temperature
T_J	junction temperature
t	time
t_{coll}	time interval between collisions
t_o	response time of field-effect transistor
t_{off}	turn-off time
t_{tr}	transit time
u	gas velocity
U	velocity of undisturbed gas
U	net rate of carrier recombination-generation in the bulk (per unit time and volume)
U_s	net rate of carrier recombination-generation at a surface (per unit time and area)
$\bar{v}_{\text{drift}}$	average drift velocity
v_{th}	thermal velocity of carriers
V	film growth rate
V	voltage
$\bar{V}_B$	average base voltage drop
V_D	drain voltage
$V_{D\text{sat}}$	drain voltage at the onset of saturation

V_F	forward bias
V_{FB}	flat-band voltage
V_G	gate voltage
V_J	applied junction voltage
V_o	voltage drop across oxide
V_P	polarizing voltage
V_R	reverse bias
V_T	turn-on or turn-off voltage
W	depletion region width of metallurgical junction
W_B	base width
W_c^*	width of the depletion region within the cylindrical region, at breakdown
W_E	emitter depth
W_{EB}	width of the emitter-base junction space-charge region
W_{epi}	width of the lowly doped region in an epitaxial device
W_o	depletion region width in equilibrium
W_s	sample thickness
x_d	width of surface depletion region
$x_{d\max}$	maximum width of surface depletion region
$x_{d\max,o}$	maximum width of surface depletion region, in equilibrium
x_j	junction depth
x_o	oxide (insulator) thickness
Y	mole fraction
Z	channel width
α	common-base current gain
α_R	common-base current gain in the reverse direction
α_T	transport factor
β	common-emitter current gain
γ	emitter efficiency
δ	average boundary layer thickness
ϵ_0	permittivity of free space (8.86×10^{-14} f/cm or 55.4 e/Vμ)
κ	thermal diffusivity
μ	viscosity
μ	mobility
μ_n	electron mobility
μ_p	hole mobility

List of Symbols

ρ	density
ρ	resistivity
ρ	space-charge density
σ	conductivity
σ	capture cross-section
τ	offset time in the general relationship for the oxidation of silicon
τ_o	effective lifetime within a reverse-biased depletion region
τ_p	lifetime of holes in an n-type semiconductor
τ_n	lifetime of electrons in a p-type semiconductor
ϕ	electrostatic potential
ϕ_B	built-in voltage of metallurgical p-n junction
ϕ_F	Fermi potential
ϕ_s	surface potential, i.e., total potential variation across surface space-charge region
ϕ_T	total potential variation across space-charge region
Φ_M	metal work function
Φ_{MS}	metal-semiconductor work function difference
Φ_S	semiconductor work function

**Physics and Technology of
Semiconductor Devices**

Introduction:
The Planar Technology

The planar technology and semiconductor devices made by this technology were first described in 1960.[1] The planar technology has since become the principal method of fabricating semiconductor devices and integrated circuits, and has strongly contributed to the rapidity with which semiconductor devices have displaced older types of electronic components and penetrated into entirely new electronic applications.

Like most important technological advances, the planar technology evolved from several generations of earlier ones. This evolution, as well as the essentials of the planar technology, are best illustrated by considering it in comparison with two of the most important earlier semiconductor device-fabrication techniques, the *grown junction method* and the *alloy junction method*. These are illustrated in the figure.

In the *grown junction method*,[2] a semiconductor crystal is grown out of a melt of the semiconductor which is doped a certain type (taken to be *p*-type in this example). At some point in the growth process, the doping concentration in the melt is suddenly changed; for instance, by dropping a pill containing donor-type impurities into the melt. As a result, the rest of the crystal will be grown *n*-type. When the growth is completed, the crystal is sectioned into little bars containing the *p-n* junction as indicated by the dashed lines.

This method was extremely important in the first years following the invention of the junction transistor. For instance, it yielded the first diodes with which the theory of the current-voltage characteristics of *p-n* junctions was verified. The grown junction method, however, was not as suitable for mass production as another method developed in the early years of semiconductor device technology, the alloy junction method.

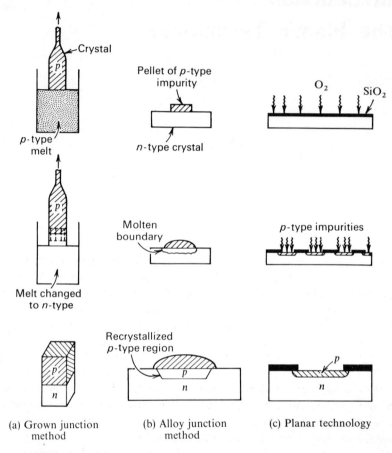

(a) Grown junction method
(b) Alloy junction method
(c) Planar technology

In the *alloy junction method*,[3] a pellet containing acceptor-type impurities (in this example) is placed upon a wafer of *n*-type semiconductor. The wafer and the pellet are then heated to a temperature high enough so that the pellet fuses or alloys into the semiconductor wafer. When the crystal is cooled, a recrystallized region which is saturated with acceptor-type impurities forms under the pellet. Thus a *p-n* junction results.

This method has been—and still is—employed with great success in the mass production of diodes and transistors, mainly made of germanium.

The Planar Technology

However, as semiconductor devices reached out for new applications, demands on their performance greatly increased. These increasing demands soon demonstrated the inherent limitations of the alloy junction method. For instance, we shall see later that the most important factor determining the performance of junction transistors is the distance separating two junctions. In the alloy process, the location of junctions was determined by the extent to which the recrystallized regions penetrated into the semiconductor. However, this penetration was always difficult to control.

The search for a method which gives superior control of the location of *p-n* junctions led to the development of *diffused junctions*.[4] Diffused junctions are formed in a manner similar to the alloying process in that the surface of the wafer is exposed to a source of a high concentration of opposite-type impurities contained, for instance, in a gas. However, no phase-formation takes place in this case; the impurities penetrate the semiconductor crystal by solid-state diffusion which can be controlled to a very precise degree.

With the additional discovery that a thin layer of silicon dioxide can effectively *mask* against the diffusion of most important acceptor and donor impurities,[5] a new degree of precision was added in controlling the geometry of diffused-junction semiconductor devices; the device geometry now could be delineated by covering the semiconductor with silicon dioxide, and then exposing the semiconductor to the diffusing impurities only in selected areas, defined by photolithography, where the oxide layer has been removed.

The *planar technology*, schematically illustrated in the last figure, combines the advantages of junction formation by solid-state diffusion and the masking property of silicon dioxide for precise definition of device geometry. It also makes use of the very important fact that the electrical characteristics of a silicon surface covered with an oxide layer are superior to those of a bare surface.

Because of this last feature, the sensitivity of semiconductor devices to their environments has been greatly reduced with a corresponding increase in the reproducibility and stability of device characteristics. Recently, thermally oxidized silicon structures have even made the construction of surface field-effect devices feasible for the first time since the conception of such devices some thirty years ago. Most importantly, the planar technology has led to the development of integrated circuits which, although only a few years old, show indications of influencing electronics—technology, design, and application—possibly to as large an extent as the advent of semiconductor devices did ten years earlier.

READING REFERENCES

The earlier methods of junction formation are summarized (with many references) by W. C. Dunlap, Section 7 in *Handbook of Semiconductor Electronics*, 2nd ed., L. P. Hunter, Editor, McGraw-Hill Book Co., 1962.

A detailed description of the planar technology and its application to integrated circuits is given by G. E. Moore, Chapter 5 in *Microelectronics*, E. Keonjian, Editor, McGraw-Hill Book Co., 1963.

REFERENCES CITED

1. J. A. Hoerni, "Planar Silicon Transistors and Diodes," *IRE Electron Devices Meeting*, Washington, D.C. (1960).
2. G. K. Teal, M. Sparks, and E. Buehler, "Growth of Germanium Single Crystals Containing *P-N* Junctions," *Phys. Rev.*, **81,** 637 (1951).
3. R. N. Hall and W. C. Dunlap, "*P-N* Junctions Prepared by Impurity Diffusion," *Phys. Rev.*, **80,** 467 (1950).
4. M. Tanenbaum and D. E. Thomas, "Diffused Emitter and Base Silicon Transistors," *Bell System Tech. J.*, **35,** 1 (1956); C. A. Lee, "A High Frequency Diffused Base Germanium Transistor," *Bell Systems Tech. J.*, **35,** 23 (1956).
5. C. J. Frosch and L. Derrick, "Surface Protection and Selective Masking During Diffusion in Silicon," *J. Electrochem. Soc.*, **104,** 547 (1957).

PART I

SOLID-STATE TECHNOLOGY

- **Vapor-Phase Growth**
- **Thermal Oxidation**
- **Solid-State Diffusion**

- KINETICS OF GROWTH
- GAS-PHASE MASS TRANSFER
- SOME PROPERTIES OF GASES

1
Vapor-Phase Growth

Vapor-phase growth techniques are employed in semiconductor technology for the deposition of metals (e.g., aluminum), insulators (e.g., SiO_2), and semiconductors (e.g., silicon). Of these processes, the most important one from the standpoint of device fabrication is the growth of *single crystal* semiconductor films upon single crystal substrates of the same semiconductor. Such growth is called *epitaxial* (Greek for "arranged upon").

The importance of epitaxial growth in semiconductor device technology is due to the ease with which the impurity concentration in the film can be adjusted independently of the impurities within the substrate by controlling their concentration in a gas. Thus epitaxial growth can be used to form *p-n* junctions between the epitaxial film and the substrate. More importantly, it can be used to grow films of relatively low impurity concentration upon substrates which contain the same type of impurity in much higher concentrations. In this manner the series resistance associated with the substrate can be reduced without otherwise changing the characteristics of semiconductor devices.

Because the epitaxial growth of semiconductors is so important, it has been studied in great detail. There are several ways in which such growth can be performed. In this chapter we study the method which is most widely used in the epitaxial growth of silicon: the vapor-phase reduction of silicon tetrachloride. (For a review of other types of vapor-phase growth reactions, see the Reading References listed at the end of this chapter.)

Two types of reactors that have been employed in the epitaxial growth of silicon are illustrated in Figure 1.1. Figure 1.1a shows the *vertical reactor* used in the pioneering work of Theuerer.[1] Figure 1.1b shows the *horizontal reactor* used in a more recent investigation by Shepherd.[2] The latter is typical of present-day industrial reactors in which a film is grown simultaneously on many wafers. In both reactors, a stream of hydrogen containing a certain concentration of silicon tetrachloride enters the reactor. The gas flows past silicon wafers resting on a susceptor which, in turn, is heated by inductive coupling to radio-frequency induction coils.

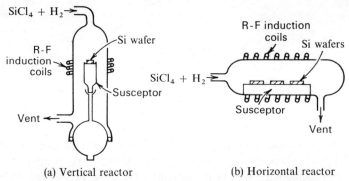

Fig. 1.1 Reactors employed in the epitaxial growth of silicon.

The reactor itself has quartz or glass walls. Since induction heating is employed, the walls remain cold during the growth process. This eliminates deposition on the reactor walls and results in minimum contamination from the walls.

While vapor-phase deposition of materials can take place at relatively low temperatures, epitaxial growth itself requires relatively high deposition temperatures, typically above 1000°C in the case of silicon. Epitaxial growth needs high temperatures because the deposited atoms must find their proper position within the crystal lattice in order to result in a single crystal film. As the temperature of deposition is reduced, the mobility of the deposited atoms decreases, and the resulting film becomes more and more defective, eventually losing its crystalline nature.

The overall reaction which results in the growth of silicon films is

$$SiCl_4 + 2H_2 \rightleftharpoons Si(solid) + 4HCl.$$

This reaction is reversible; i.e., it can take place in either direction. If the carrier gas entering the reactor contains HCl, removal or *etching* rather than growth of silicon will take place.

Vapor-Phase Growth

It is also known that an additional competing reaction takes place simultaneously with the growth reaction, given by

$$SiCl_4 + Si(solid) \rightleftharpoons 2SiCl_2.$$

As a result, if the $SiCl_4$ concentration is very high, etching of the silicon will take place even in the absence of a significant concentration of HCl in the incoming gas stream. This is illustrated in Figure 1.2 which shows

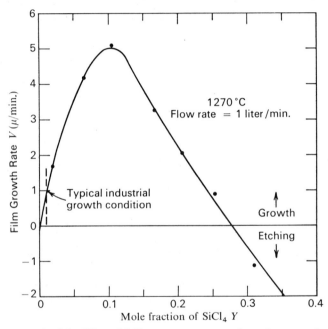

Fig. 1.2 Effect of $SiCl_4$ concentration on silicon deposition.[1]

the effect of the concentration of $SiCl_4$ in the gas on the reaction. (Concentrations in gases are most conveniently given in terms of *mole fractions*. The mole fraction Y is defined as the ratio of the number of molecules of a given species to the total number of molecules in the gas.) It is evident that initially the deposition rate increases with increasing concentration of $SiCl_4$. As the concentration of $SiCl_4$ is increased, a maximum growth rate is reached at a mole fraction of about 0.1. Further increase in concentration brings about a *decrease* in the growth rate and eventually even results in etching of the silicon surface.

Silicon is usually grown in the low concentration region, as indicated in Figure 1.2, with a typical growth rate of $\sim 1\ \mu/\text{min}$. In this region the dependence of film growth rate V on the mole fraction of $SiCl_4$ in the gas mixture Y is approximately linear.

1.1 KINETICS OF GROWTH

We shall now study the kinetics of film growth on the basis of a very simple model.[3] The essentials of this model are depicted in Figure 1.3 where we show the concentration distribution of the silicon tetrachloride in the gas and we indicate the flux of the silicon tetrachloride from the

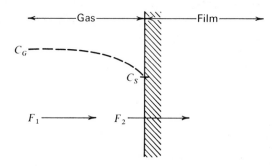

Fig. 1.3 Model of the growth process. Direction of gas flow is normal to plane of paper.

bulk of the gas to the surface of the growing film F_1, and the flux corresponding to the silicon tetrachloride consumed in the film-growth reaction F_2. (Flux is defined as the number of atoms or molecules crossing a unit area in a unit time.)

We approximate the flux F_1 by the linear formula

$$F_1 = h_G(C_G - C_S) \tag{1.1}$$

where C_G and C_S are the concentrations of the $SiCl_4$ (molecules per cubic centimeter) in the bulk of the gas and at the surface, respectively, and h_G is the *gas-phase mass-transfer coefficient*. The flux consumed by the chemical reaction taking place at the surface of the growing film F_2 is approximated by the formula

$$F_2 = k_S C_S \tag{1.2}$$

where k_S is the *chemical surface-reaction rate constant*. These linear approximations are analogous to Ohm's law: they describe a flux as being proportional to a *driving force*. In the case of mass transfer, the driving force is a concentration difference; in the case of a *first-order* chemical

Kinetics of Growth

reaction, which we assume this one to be, it is the concentration of the reacting species.†

In steady state $F_1 = F_2 = F$. Using this condition, we can solve the above two equations to obtain the surface concentration of the silicon tetrachloride at the gas-film interface,

$$C_S = \frac{C_G}{1 + k_S/h_G}. \tag{1.3}$$

This formula shows that the surface concentration will go to zero if $h_G \ll k_S$. This condition is commonly referred to as *mass-transfer control*. On the other hand, if $h_G \gg k_S$, the surface concentration approaches C_G. This condition is commonly referred to as *surface-reaction control*.

We can now readily express the growth rate of the silicon film by writing

$$V = \frac{F}{N_1} = \frac{k_S h_G}{k_S + h_G} \frac{C_G}{N_1} \tag{1.4}$$

where N_1 is the number of silicon atoms incorporated into a unit volume of the film. Its value for silicon is 5.0×10^{22} cm^{-3}. Noting that $C_G = YC_T$ where C_T is the total number of molecules per cubic centimeter in the gas, we get the expression for the growth rate,

$$V = \frac{k_S h_G}{k_S + h_G} \frac{C_T}{N_1} Y. \tag{1.5}$$

Note the following features of this equation. First, it predicts that the film growth rate V is proportional to the mole fraction Y of the reacting species. As we have seen earlier, this is in agreement with the experimental observations for small values of Y such as are encountered in usual practice. Second, the growth rate at a given mole fraction is determined by the *smaller* of h_G or k_S. This corresponds to the two limiting cases of mass-transfer controlled and surface-reaction controlled conditions. In these limiting cases the growth rate will be given either by

$$V \doteq \frac{C_T}{N_1} k_S Y \quad \text{[surface-reaction control]} \tag{1.6}$$

or by

$$V \doteq \frac{C_T}{N_1} h_G Y \quad \text{[mass-transfer control]}. \tag{1.7}$$

† Strictly speaking, we should include in our consideration the flux of the reaction product HCl from the surface back to the bulk of the gas. In this treatment we neglect this flux for simplicity. This is equivalent to the assumption that the mass-transfer coefficient of the reaction product is much larger than that of SiCl$_4$.

The temperature dependence of the film growth rate observed experimentally by Shepherd[2] is shown by the points in Figure 1.4. It is evident that at low temperatures the growth rate follows an exponential law, $V \propto e^{-E_a/kT}$. The *activation energy* E_a is about 1.9 ev. Theuerer[1] also

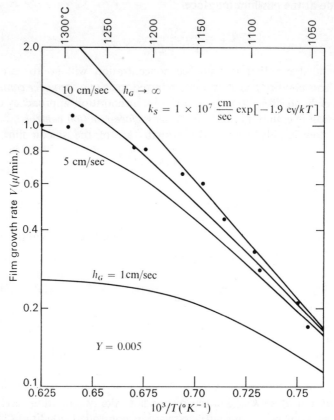

Fig. 1.4 Temperature dependence of the film-growth rate. Points represent experimental data,[2] curves were calculated based on Equation 1.5.

observed an exponential temperature dependence in this range corresponding to an activation energy of 1.6 ev. At high temperatures the growth rate levels off, and becomes relatively temperature insensitive.

Since chemical reaction rate constants generally follow an exponential temperature dependence while mass-transfer coefficients are relatively insensitive to variations in the temperature, the simple model explains the general features of the temperature-dependence data shown in Figure 1.4. By fitting Equation 1.5 to these data, we can obtain theoretical curves such as shown in that figure. The best fit seems to be obtained for the values of

Gas-Phase Mass Transfer

$h_G \simeq 5$ to 10 cm/sec and $k_S = 10^7 \text{ cm/sec } e^{-1.9 \text{ ev}/kT}$. Numbers of the same order of magnitude can be extracted from Theuerer's data obtained in the vertical reactor.

This model of the growth process is quite oversimplified. It does not consider the flux of the reaction product; the linear approximation that describes the surface reaction is, as we have discussed earlier, valid only for low values of Y. Also, the temperature gradient present in a cold-wall reactor is extremely steep. Because of this, the properties of the gas will vary radically between the heated substrate and the cold glass wall. Nevertheless, this simple model still predicts the overall features observed experimentally. It describes the two regions of the growth process—the mass-transfer and the surface-reaction controlled regions—and yields at least an order of magnitude estimate of both the chemical surface-reaction rate constant k_S and the gas-phase mass-transfer coefficient h_G from the growth-rate data.

In order to obtain a good quality epitaxial film, growth temperatures have to be relatively high. As a result, most industrial epitaxial processes take place in the mass-transfer controlled region, so in the next section we consider the mechanisms which determine gas-phase mass-transfer coefficients.

1.2 GAS-PHASE MASS TRANSFER

In Section 1.1 we approximated the flux from the bulk of the gas to the surface of the solid by the formula

$$F_1 = h_G(C_G - C_S).$$

We now consider the factors which determine h_G.

a. Stagnant-film Model[4]

The simplest possible picture of the mass-transfer process is shown in Figure 1.5. In this picture we split the gas phase into two regions: one which is well mixed and which flows past the solid surface with a uniform velocity U, and another, a stagnant film of thickness δ next to the solid. Transport of the active species across this stagnant film proceeds by diffusion alone. Thus the flux F_1 can be immediately written down as

$$F_1 = D_G \frac{C_G - C_S}{\delta} \quad (1.8)$$

where D_G is the *diffusivity* of the active species in the gas. This results in a formula for the gas-phase mass-transfer coefficient,

$$h_G = \frac{D_G}{\delta}. \tag{1.9}$$

Although this stagnant-film model is quite arbitrary and fictitious, it has been successfully applied to many problems involving gas-phase mass

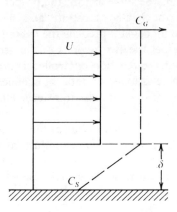

Fig. 1.5 The "stagnant-film" model of gas-phase mass transfer.

transfer since the 1930's. However, the thickness of the stagnant film being an arbitrary quantity, it has to be determined experimentally for any given set of conditions.

b. Boundary Layer Theory[5]

Fluid mechanics can provide a more realistic and useful estimate of the mass-transfer coefficient h_G. Consider the geometrically simplest problem: a fluid flowing parallel to a flat plate of length L. We assume that the extent of the flowing fluid is infinitely large or, in other words, that the fluid is unbounded. Far away from the plate the fluid flows with a uniform velocity U as shown in Figure 1.6. Right next to the plate the velocity of the fluid is zero. The reason for this is as follows. The frictional force per unit area along the x-direction acting on a fluid element next to the plate is given by

$$F_{\text{friction}} = \mu \frac{\partial u}{\partial y} \tag{1.10}$$

where μ is the *coefficient of viscosity* or simply *viscosity*. The plate is stationary, i.e., its velocity is zero. If right next to the plate the fluid

Gas-Phase Mass Transfer

possessed a finite velocity, the velocity gradient there would be infinitely large. Then according to Equation 1.10 an infinitely large frictional force would act on the fluid, immediately bringing its velocity to zero. Thus the velocity of the fluid right next to a solid body must clearly be zero in order to avoid the existence of infinitely large velocity gradients.

The boundary condition of zero velocity right next to the plate disturbs the velocity distribution. As the fluid moves along the plate, this disturbance will spread further and further into the bulk of the fluid. To calculate the thickness of the disturbed region δ, we recall that *it is the friction at the wall that causes the deceleration of the fluid.*

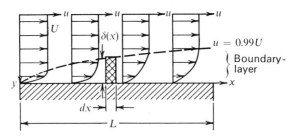

Fig. 1.6 Development of a boundary layer in flow past a flat plate.

According to Newton's second law,

$$F = ma. \qquad (1.11)$$

Consider now the shaded element shown in Figure 1.6, and take it to represent a volume element whose size in the direction normal to the paper is unity.

The principal force acting on this element is $F_{\text{friction}}\, dx$.

The acceleration of this element is

$$\frac{du}{dt} = \frac{du}{dx}\frac{dx}{dt} = \frac{du}{dx}u.$$

The mass of this element is $\rho\delta(x)\, dx$. Here ρ denotes the density. If we combine the above with Newton's second law, we get

$$\mu\frac{\partial u}{\partial y} = \rho\delta(x) u \frac{du}{dx}. \qquad (1.12)$$

Replacing the differentials by the respective differences results in

$$\mu\frac{U}{\delta(x)} \simeq \rho\delta(x) U \frac{U}{x} \qquad (1.13)$$

which can be rearranged to lead to an expression giving an estimate of the size of the region of disturbed velocity commonly referred to as the *boundary layer thickness*:

$$\delta(x) \cong \sqrt{\frac{\mu x}{\rho U}}. \quad (1.14)$$

More exact calculations by Blasius[5] in the 1930's, based on the solution of the complete equations of motion for a fluid, led to a formula which differs from Equation 1.14 only by a numerical coefficient whose value is between $\frac{2}{3}$ and 5, depending on the precise definition of δ. (In Figure 1.6 the dashed line indicates the location of those points where the velocity reaches 99% of the velocity of the free stream U. This is one way of defining δ.)

The average boundary layer thickness $\bar{\delta}$ over the whole plate is given by

$$\bar{\delta} \equiv \frac{1}{L}\int_0^L \delta(x)\, dx = \tfrac{2}{3}L\sqrt{\mu/\rho U L} \quad (1.15)$$

or

$$\bar{\delta} = \frac{2}{3}\frac{L}{\sqrt{\mathrm{Re}_L}}. \quad (1.16)$$

This expression gives the average boundary layer thickness in terms of the *Reynolds number*,

$$\mathrm{Re}_L \equiv \frac{\rho U L}{\mu}.$$

The Reynolds number is an extremely important dimensionless number in fluid dynamics. It represents the ratio of the magnitude of inertial effects to viscous effects in fluid motion. Thus large Reynolds numbers imply small viscous effects, and vice versa.

Now, if we take the average boundary layer thickness $\bar{\delta}$ for the thickness of the stagnant film δ in Equation 1.9, we get an expression for the mass-transfer coefficient h_G:

$$h_G = \frac{D_G}{\bar{\delta}} = \frac{3}{2}\frac{D_G}{L}\sqrt{\mathrm{Re}_L} \quad (1.17)$$

which can be rearranged into the dimensionless equation

$$\frac{h_G L}{D_G} = \tfrac{3}{2}\sqrt{\mathrm{Re}_L}. \quad (1.18)$$

This can be compared with the result of a more exact calculation by Pohlhausen:[6,7]

$$\frac{h_G L}{D_G} = \tfrac{2}{3}\sqrt{\mathrm{Re}_L}\sqrt[3]{\mu/\rho D_G} \quad (1.19)$$

Gas-Phase Mass Transfer

where $\mu/\rho D_G$, a dimensionless group called the *Schmidt number*, has values between 0.6 and 0.8 for most gases practically independently of temperature. Thus the numerical factor by which our simple derivation deviated from the result of the more exact treatment is of the order of unity.

We can now estimate the mass-transfer coefficient for an epitaxial reactor such as employed in Theuerer's[1] and Shepherd's[2] work. Typical velocities are 10 to 30 cm/sec, leading to values of the Reynolds number of $\sim$20. Using Equation 1.19, this leads to $h_G \simeq 5$ cm/sec. This is in

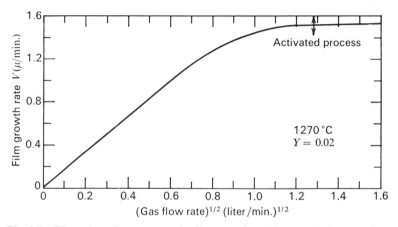

Fig. 1.7 Effect of gas-flow rate on the film growth rate in a vertical reactor.[1]

order-of-magnitude agreement with the value we had obtained by fitting our model to both Shepherd's and Theuerer's experimental data. (The estimation of viscosity, density, and diffusivity are discussed in Section 1.3.)

Theuerer's data of film growth rate as a function of the square root of gas-flow rate in the vertical reactor are shown in Figure 1.7. It appears that at low gas-flow rates the film growth rate V increases approximately in proportion to the square root of the gas-flow velocity. At high flow rates, V reaches a maximum; in this region V depends exponentially on temperature, indicating that the growth rate has become surface-reaction limited.

In contrast, Shepherd found no effect of the gas-flow rate on the mass-transfer coefficient in the horizontal reactor. To explain this discrepancy, we must recall that the length of the effective flat plate (the susceptor) in Shepherd's horizontal reactor is considerably longer than in Theuerer's vertical reactor. In fact, the average boundary-layer thickness for Shepherd's reactor can be estimated to be of the order of 2 to 3 cm, which is as large as the radius of the reactor. Thus, evidently, the unbounded fluid approximation cannot be expected to yield reasonable results.

In such a case we may use another simple approximation based on Graetz's theory,[8] developed in 1885 for heat transfer in a circular tube. Because the transport of heat and of mass are described by the same equations,[7] Graetz's theory can be readily adapted to mass-transfer considerations as shown in Figure 1.8. It is evident that for both a parabolic

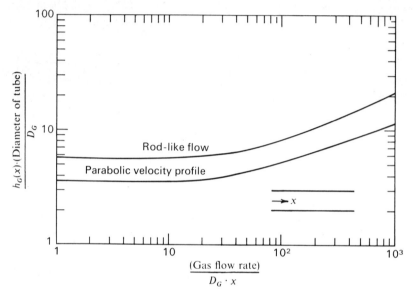

Fig. 1.8 The Graetz-solution for mass transfer from the wall to the fluid in a circular tube.[8]

velocity profile and for a rod-like flow (uniform velocity profile) the mass-transfer coefficient for low gas-flow rates becomes independent of the flow rate. The conditions in Shepherd's reactor correspond to an abscissa of the order of 1. This leads to a mass-transfer coefficient of the same order as the boundary layer considerations did. However, this theory predicts the complete absence of a gas flow-rate dependence of the mass-transfer coefficient, in agreement with Shepherd's observation.

1.3 SOME PROPERTIES OF GASES

We now briefly review how certain properties of gases, which are needed in mass-transfer calculations, are obtained.

Some Properties of Gases

a. Ideal Gas Law

The ideal gas law states that

$$pV_{\text{molar}} = RT = (\text{Avogadro's number}) \cdot kT \qquad (1.20)$$

where the gas constant R is conveniently given for such calculations by

$$R = 82.1 \frac{\text{cm}^3 \text{ atm}}{(\text{g-mole})°K},$$

and the Boltzmann constant k is conveniently given by

$$k = 1.37 \times 10^{-22} \frac{\text{cm}^3 \text{ atm}}{(\text{molecule})°K}.$$

This can be rearranged to give the concentration in units of molecules per cubic centimeter as

$$C = \frac{p}{kT}. \qquad (1.21)$$

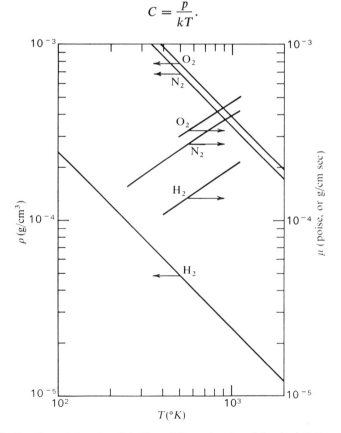

Fig. 1.9 Density and viscosity of H_2, N_2, and O_2 as a function of the absolute temperature.

The ideal gas law can also be rearranged to give directly the density of a gas,

$$\rho = (\text{molecular weight}) \frac{p}{RT}. \qquad (1.22)$$

The densities of three gases—oxygen, nitrogen, and hydrogen—are shown in Figure 1.9 as functions of the absolute temperature.

b. Transport Properties

The estimation of viscosities and diffusivities of gases is treated thoroughly elsewhere.[9] To provide a feeling for the orders of magnitude involved, we show the viscosity of the same three gases as a function of absolute temperature in Figure 1.9.

Diffusivities of gases at room temperature generally range between 0.1 and 1 cm²/sec.[9] Gaseous diffusivities generally increase with absolute temperature as T^m, with m between 1.5 and 2.

READING REFERENCES

A general discussion of vapor-phase growth is given by C. P. Powell, J. H. Oxley, and J. M. Blocher, *Vapor Deposition*, Wiley, 1966.

The epitaxial growth of semiconductors is discussed in a series of papers in the December 1963 issue of *RCA Review*; and in Volume 9 of a series of review-reports on "Integrated Silicon Device Technology" by the Research Triangle Institute, ASD-TDR-63-316 (1965).

Gas-phase mass transfer is treated by R. E. Treybal, *Mass-Transfer Operations*, McGraw-Hill Book Co., 1955; and by R. B. Bird, W. E. Stewart, and E. N. Lightfoot, *Transport Phenomena*, Wiley, 1960.

REFERENCES CITED

1. H. C. Theuerer, "Epitaxial Silicon Films by the Hydrogen Reduction of SiCl$_4$," *J. Electrochem. Soc.*, **108**, 649 (1961).

2. W. H. Shepherd, "Vapor Phase Deposition and Etching of Silicon," *J. Electrochem. Soc.*, **112**, 988 (1965).

3. A. S. Grove, "Mass Transfer in Semiconductor Technology," *Ind. & Eng. Chem.*, **58**, 48 (1966).

4. R. E. Treybal, *Mass-Transfer Operations*, McGraw-Hill Book Co., 1955, Chapter 3.

5. H. Schlichting, *Boundary Layer Theory*, 4th ed., McGraw-Hill Book Co., 1960, Chapter 7.

Problems

6. *Ibid.*, Chapter 14.
7. R. B. Bird, W. E. Stewart, and E. N. Lightfoot, *Transport Phenomena*, Wiley, 1960, Chapter 21.
8. W. H. McAdams, *Heat Transmission*, 3rd ed., McGraw-Hill Book Co., 1954, Chapter 9.
9. R. C. Reid and T. K. Sherwood, *The Properties of Gases and Liquids*, McGraw-Hill Book Co., 1958.

PROBLEMS

1.1 Exponential temperature dependence is alternately represented by the form $e^{-E_a/kT}$ or $e^{-\Delta H/RT}$, where E_a is the value of the activation energy in electron volts, and ΔH is its value in Kcal/mole. Show that if $E_a = 1$ ev, the corresponding value of ΔH is 23 Kcal/mole.

1.2 Calculate the time required to grow a monolayer of silicon epitaxially under the conditions of Figure 1.4, at 1200°C.

1.3 Derive the expression for the film-growth rate V if deposition takes place through the chemical reaction,

$$SiX \rightleftharpoons Si(solid) + X,$$

in terms of the rate constants of the forward and reverse chemical reactions k_s and k_{sr}, and the gas-phase mass-transfer coefficients of SiX and of X, h_G and h_{GX}. Under what conditions does the resulting expression reduce to Equation 1.5?

1.4 A tubular flow reactor is 2" in diameter. The flow meter indicates a gas-flow rate of 1 liter/min. The entering gas mixture is 98% H_2, 2% $SiCl_4$ by volume. The reactor is heated to 1200°C. Calculate:
(a) The Reynolds number based on the wafer size.
(b) The average boundary layer thickness over the first wafer.
(c) The concentration of $SiCl_4$ (molecules/cm³) in the entering gas.
(d) The flux of $SiCl_4$ molecules to the surface.
(e) The film growth rate, assuming the growth is mass-transfer limited.
(f) The concentration of $SiCl_4$ in the vent gas, assuming that there are 20 wafers in the reactor, and that the film growth rate is the same on all wafers and is that calculated in (e).

- **KINETICS OF OXIDE GROWTH**
- **SPACE-CHARGE EFFECTS**

2

Thermal Oxidation

The growth of a thin layer of silicon dioxide on a silicon wafer is a basic feature of the planar technology. Both the precise control of the thickness of the oxide layer and a knowledge of the kinetics of the oxidation process are therefore of obvious importance in the fabrication of planar devices. In addition, phenomena involved in the oxidation process play an important role in determining the electrical characteristics of planar silicon devices.

The mechanism of oxidation of silicon is also of general scientific interest. The oxidation of certain metals such as copper and aluminum has been studied extensively in the past in connection with corrosion problems. Silicon now provides another example of an element whose oxidation kinetics and mechanism are relatively well understood.

Silicon dioxide (SiO_2) layers on silicon can be formed by various methods. For example, they can be deposited through a vapor-phase reaction,[1] or they can be formed by electrochemical oxidation (anodization)[2] or by a plasma reaction.[3] In industrial practice, silicon dioxide layers are most frequently formed by the *thermal oxidation* of silicon[4] through the chemical reaction

$$Si(solid) + O_2 \rightarrow SiO_2(solid)$$

or

$$Si(solid) + 2H_2O \rightarrow SiO_2(solid) + 2H_2.$$

Kinetics of Oxide Growth

Thus some of the silicon is used up in the growth of an oxide film. It can be shown from the densities and molecular weights of silicon and silicon dioxide† that in the growth of an oxide film whose thickness is x_o, a layer of silicon $0.45x_o$ thick is consumed.

In this chapter we study the thermal oxidation of silicon in some detail. The thermal oxidation of silicon is performed in oxidation furnaces or reactors of the type illustrated schematically in Figure 2.1. The reactor itself is tubular, usually made of quartz or glass. It is heated by a resistance furnace to temperatures in the vicinity of 1000°C. A gas containing the oxidizing medium (oxygen or water vapor) flows through the reactor and

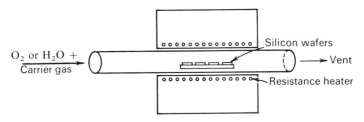

Fig. 2.1 Oxidation furnace.

past the silicon wafers. Gas-flow velocities are typically of the order of 1 cm/sec. At typical oxidation temperatures such a flow velocity corresponds to a Reynolds number Re $\approx$ 10 to 20. (See Chapter 1.)

After the oxidation step is completed, the oxide thickness can be determined by either precise measurement of the weight gain of the silicon sample or by optical techniques. These two methods have been shown to be in excellent agreement.[5]

2.1 KINETICS OF OXIDE GROWTH

We shall study the kinetics of the oxidation process on the basis of the simple model[6] illustrated in Figure 2.2.

It has been demonstrated by the use of radioactive tracers[4] and also by other means[7] that oxidation of silicon proceeds by the inward motion of the oxidizing species through the oxide layer rather than by the opposite process of the outward motion of silicon to the outer surface of the oxide. (This forms an interesting contrast to the case of copper whose oxidation proceeds by the outward motion of the metallic ion,[8] and also to the case of anodic oxidation of silicon where silicon moves outward.[9])

† Some important properties of SiO_2 are listed in Table 4.1.

For the oxidizing species to reach the silicon surface it must go through three consecutive steps:

1. It must be transported from the bulk of the gas to the oxide-gas interface.
2. It must diffuse across the oxide layer already present.
3. It must react at the silicon surface.

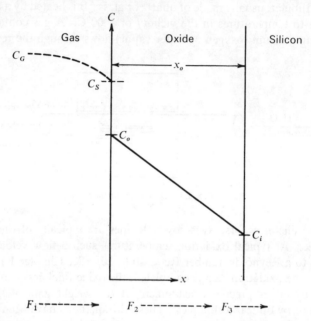

Fig. 2.2 Model for the thermal oxidation of silicon.[6] Direction of gas flow is normal to plane of paper.

The three fluxes corresponding to the three steps are equal in steady-state oxidation. They will be approximated as follows. As in Chapter 1, a linear approximation to the gas-phase flux F_1 is obtained by taking the flux of oxidant from the bulk of the gas to the oxide-gas interface to be proportional to the difference between the concentration of the oxidant in the bulk of the gas C_G and the concentration of the oxidant right next to the oxide surface C_S,

$$F_1 = h_G(C_G - C_S). \tag{2.1}$$

Here h_G is the mass-transfer coefficient which we discussed in the previous chapter.

We now assume *Henry's law* to hold. Henry's law states that, in equilibrium, the concentration of a species within a solid is proportional to the

Kinetics of Oxide Growth

partial pressure of that species in the surrounding gas. Thus we assume that the concentration at the outer surface of the oxide C_o is proportional to the partial pressure of the oxidant right next to the oxide surface p_S, i.e., $C_o = Hp_S$, where H is the Henry's law constant. Furthermore, we denote the equilibrium concentration in the oxide, i.e., the concentration which would be in equilibrium with the partial pressure in the *bulk* of the gas p_G by the symbol C^*, i.e., $C^* = Hp_G$.

If we recall that according to the ideal gas law (see Chapter 1) $C_G = p_G/kT$, and $C_S = p_S/kT$, we can rewrite Equation 2.1 as

$$F_1 = h(C^* - C_o) \tag{2.2}$$

where h is the gas-phase mass-transfer coefficient *in terms of concentrations in the solid*, given by $h = h_G/HkT$.

The flux across the oxide F_2 is taken to be a diffusive flux,

$$F_2 = D\frac{C_o - C_i}{x_o}, \tag{2.3}$$

where D is the diffusivity of the oxidizing species in the oxide layer.

Finally, the rate of the reaction taking place at the oxide-silicon interface is assumed to be proportional to the concentration of the oxidant at this interface. Thus,

$$F_3 = k_S C_i \tag{2.4}$$

where k_S is the chemical surface-reaction rate constant for oxidation.†

Using the condition of steady state, $F = F_1 = F_2 = F_3$, after some algebra, we get the following expressions for the concentrations of the oxidant at the oxide-silicon interface C_i and at the gas-oxide interface C_o,

$$C_i = \frac{C^*}{1 + \dfrac{k_S}{h} + \dfrac{k_S x_o}{D}} \tag{2.5}$$

and

$$C_o = \frac{\left(1 + \dfrac{k_S x_o}{D}\right) C^*}{1 + \dfrac{k_S}{h} + \dfrac{k_S x_o}{D}}. \tag{2.6}$$

It is interesting to consider the two limiting forms of Equations 2.5 and 2.6. When the diffusivity is very small, $C_i \to 0$ and $C_o \to C^*$. In the opposite case, when the diffusivity is very large, C_i and C_o will be equal and

---

† Note the similarity between this model and the one used in the treatment of vapor-phase growth in Chapter 1.

will be given by $C^*/(1 + k_S/h)$. These two limiting cases are called the *diffusion-controlled* and *reaction-controlled* cases, respectively. The distributions of the concentration of the oxidizing species in the oxide layer for these two limiting cases are illustrated in Figure 2.3. (We are assuming in this figure that k is much smaller than h. We shall see later that this is indeed a realistic assumption.)

In order to calculate the rate of oxide growth we will have to define one more quantity, N_1, which is the number of oxidant molecules incorporated into a unit volume of oxide. There are 2.2×10^{22} SiO_2 molecules/cm^3 in the oxide and we incorporate one O_2 molecule into each SiO_2

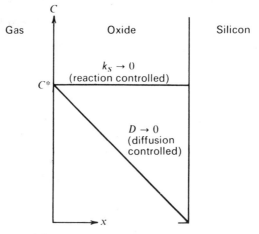

Fig. 2.3 Distribution of the oxidizing species in the oxide layer for the two limiting cases of oxidation.

molecule, whereas we incorporate two H_2O molecules into each SiO_2 molecule. Hence, N_1 for oxidation in dry oxygen will be 2.3×10^{22} cm^{-3}, whereas for oxidation in water vapor it will be twice this number.

Thus the flux of oxidant reaching the oxide-silicon interface is given by

$$N_1 \frac{dx_o}{dt} = F = \frac{k_S C^*}{1 + \frac{k_S}{h} + \frac{k_S x_o}{D}}. \qquad (2.7)$$

We solve the above differential equation subject to the initial condition, $x_o(0) = x_i$. Here x_i is the thickness of oxide layer grown in an earlier oxidation step. This general initial condition permits us to consider two or more successive oxidation steps. The quantity x_i can also be regarded as the thickness of oxide that is present at the end of an initial period of oxidation during which the assumptions involved in this simple

Kinetics of Oxide Growth

oxidation model may not have been valid. We will consider the significance of x_i in this case in more detail in Section 2.2.

The solution of the differential equation (2.7) leads to the general relationship for the oxidation of silicon,[6]

$$x_0^2 + Ax_0 = B(t + \tau) \qquad (2.8)$$

where $A \equiv 2D\left(\dfrac{1}{k_S} + \dfrac{1}{h}\right)$ (2.8a)

$$B \equiv \dfrac{2DC^*}{N_1} \qquad (2.8b)$$

and $\tau \equiv \dfrac{x_i^2 + Ax_i}{B}.$ (2.8c)

Equation 2.8 can be solved for the oxide thickness as a function of time, resulting in

$$\dfrac{x_o}{A/2} = \sqrt{1 + \dfrac{t + \tau}{A^2/4B}} - 1. \qquad (2.9)$$

This relationship is shown in Figure 2.4 along with a large number of experimental data taken by various investigators under widely varying conditions of temperature, partial pressure of oxidants, using either oxygen or water vapor as oxidizing species.

Note the two limiting cases of this general relationship. For large times, i.e., $t \gg A^2/4B$, the so-called *parabolic relationship*, $x_0^2 = Bt$ is approached. In this relationship B is referred to as the *parabolic rate constant*. For small times, i.e., $(t + \tau) \ll A^2/4B$, we obtain the *linear law*, $x_0 = (B/A)(t + \tau)$ where B/A is referred to as the *linear rate constant* and is given by

$$\dfrac{B}{A} = \dfrac{k_S h}{k_S + h} \dfrac{C^*}{N_1}. \qquad (2.10)$$

These two limiting cases are also illustrated in Figure 2.4.

As is evident from Figure 2.4, the experimental results follow the predictions of the simple model of oxidation very well over a wide range of conditions. In addition, the predicted effect of temperature and pressure on the coefficients A and B of the general relationship have all been verified experimentally. In particular, it is found that B is proportional to the partial pressure of the oxidant in the gas, indicating that the assumption of

Henry's law was indeed justified. This implies the absence of any dissociation effects at the gas-oxide interface. Thus, for oxidation both with oxygen and with water vapor, the oxidizing species moving through the oxide layer are apparently *molecular*.

The principal effect of temperature on B should be reflected in the diffusivity D_{eff}. Experimental measurements of the temperature dependence of the parabolic rate constant B are shown in Figure 2.5. It is seen

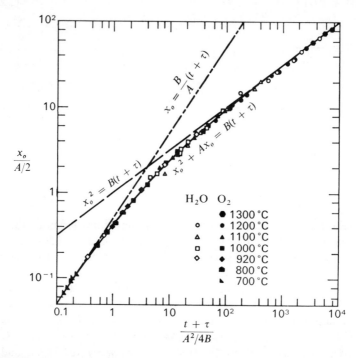

Fig. 2.4 The general relationship for silicon oxidation and its two limiting forms.[6]

that the dependence is exponential with activation energies of 0.71 ev and 1.24 ev for the cases of oxidation in oxygen and in water vapor, respectively. These activation energies are very close to the respective activation energies of the diffusivities of water vapor and oxygen through fused silica (bulk SiO_2). Using the known values of these diffusivities, we can calculate the equilibrium concentration of the oxidizing species in the oxide C^*. It is found that these numbers (5×10^{16} cm^{-3} for oxygen and 3×10^{19} cm^{-3} for water) are in good agreement with independent measurements of the solubility of oxygen and water in fused silica, at atmospheric pressure.

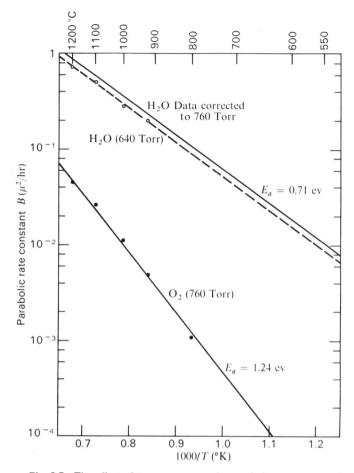

Fig. 2.5 The effect of temperature on the parabolic rate constant.[6]

The temperature dependence of the linear rate constant B/A is shown in Figure 2.6. It is evident that the linear rate constant B/A also depends on temperature exponentially with an activation energy of about 2 ev for both oxidants. The fact that the temperature dependence of B/A is exponential and that 2 ev is close to the bond-breaking energy of silicon as obtained by independent techniques,[10] indicates that the rate constant which dominates B/A is the one associated with the reaction at the oxide-silicon interface, k_S. (We can see from Equation 2.10 that of k_S and h the smaller will determine the magnitude of B/A.)

Confirmation of the fact that it is the oxide-silicon interface reaction-rate constant k_S that controls B/A, and not the gas-phase mass-transfer

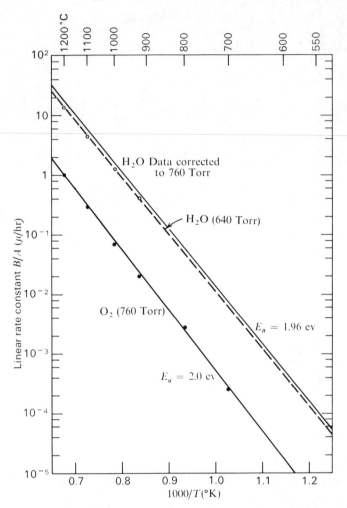

Fig. 2.6 The effect of temperature on the linear rate constant.[6]

coefficient h, is supplied by the realization that the crystallographic orientation of the single crystal silicon sample influences the magnitude of the linear rate constant B/A[11]. [All data shown in this chapter are for (111) orientation.] Also, estimates of h based on boundary layer theory, as outlined in Chapter 1, show that h is about 1,000 times larger than $k_S h/(k_S + h)$. Direct experimental evidence of the absence of a gas-phase transport limitation in thermal oxidation is supplied by the observation that a 50-fold variation in gas-flow rate has no effect on the value of B/A.[6]

The experimental observations provide a very good overall verification

Space-Charge Effects on Oxidation

of the model of oxidation discussed in this section. There remains one important unexplained observation. It is found that data obtained by oxidation in oxygen can only be brought into consistent agreement with theory if we assume a fictitious initial condition of $x_i = 200$ Å. Careful measurements of oxide growth in dry oxygen at low temperatures indicate that there is, in fact, a rapid initial phase of oxidation. This is illustrated in Figure 2.7, where the initial rapid oxidation phase is indicated as well as the

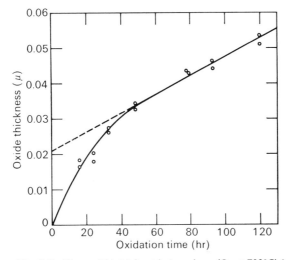

Fig. 2.7 The rapid initial oxidation phase (O_2 at 700°C).[6]

subsequent linear growth. It is evident that the linear portion of the oxide thickness versus time curve extrapolates to approximately $x_i = 200$ Å rather than to zero. This has been found consistently through all experiments with dry oxygen. However, such rapid initial phase (or finite x_i) has never been observed in oxidation in a wet ambient.

A possible explanation for this observation might be based on the role of space charges in the oxidation process.

2.2 SPACE-CHARGE EFFECTS ON OXIDATION

So far we have treated the oxidation process as if it proceeded through the motion of an uncharged oxidizing species across the oxide. However, this may not be the case.

As we have stated earlier, it is known that oxidation proceeds through the motion of the oxidizing species across the oxide, rather than through the motion of silicon. Other experiments studying the effect of an electric field applied across the oxide on the oxidation rate have indicated that the oxidizing species in oxidation in O_2 is negatively charged.[12]

We can now speculate about the reaction taking place at the gas-oxide interface. Let us suppose, for instance, that molecular oxygen from the gas gets adsorbed at the interface. Upon its entry into the oxide, it dissociates into a negatively charged oxygen molecule and a positive hole according to the reaction

$$O_2 \rightleftharpoons O_2^- + \text{hole}.$$

Both the oxygen ion and the hole begin to move across the oxide layer toward the silicon. However, the net electric current flowing across the oxide must still be zero. The hole, which has a higher mobility, moves faster. In effect, it runs ahead of the slower ion and drags the ion with it. Such a consideration is not restricted to the case of thermal oxidation of silicon. Whenever we have the coupled motion of two charged species that have different mobilities, a built-in electric field will result. This field *aids the motion of the slower of the two species.*

Cabrera and Mott[8] have considered this process in detail. Their theory predicts that so long as the oxide thickness is small—smaller than a critical distance—there will be an initial rapid oxidation phase. As the oxide grows beyond this critical thickness, its rate of growth will slow down. Estimates of this critical thickness (called extrinsic Debye length) yield 150 Å for silicon oxidation in O_2 and 5 Å for oxidation in H_2O. Since the first value is in agreement with the observed thickness below which the oxidation rate is rapid in O_2-oxidation, and the second is too small to be detectable experimentally, this mechanism may account for the experimental observations.

READING REFERENCES

A comprehensive treatment of the oxidation of metals is given by U. R. Evans, *The Corrosion and Oxidation of Metals*, Edward Arnold and Co., London 1960.

The oxidation of silicon is reviewed in Volume 7 of a series of review-reports on "Integrated Silicon Device Technology" by the Research Triangle Institute, ASD-TDR-63-316 (1965).

REFERENCES CITED

1. E. L. Jordan, "A Diffusion Mask for Germanium," *J. Electrochem. Soc.*, **108**, 478 (1961).
2. P. F. Schmidt and W. Michel, "Anodic Formation of Oxide Films on Silicon," *J. Electrochem. Soc.*, **104**, 230 (1957).

3. J. R. Ligenza, "Silicon Oxidation in an Oxygen Plasma Excited by Microwaves," *J. Appl. Phys.*, **36**, 2703 (1965).

4. M. M. Atalla, "Semiconductor Surfaces and Films; The Silicon-Silicon Dioxide System," in *Properties of Elemental and Compound Semiconductors*, H. Gatos, Ed., Vol. 5, pp. 163–181, Interscience, 1960.

5. B. E. Deal, "The Oxidation of Silicon in Dry Oxygen, Wet Oxygen, and Steam," *J. Electrochem. Soc.*, **110**, 527 (1963).

6. B. E. Deal and A. S. Grove, "General Relationship for the Thermal Oxidation of Silicon," *J. Appl. Phys.*, **36**, 3770 (1965).

7. J. R. Ligenza and W. G. Spitzer, "The Mechanisms for Silicon Oxidation in Steam and Oxygen," *Phys. Chem. Solids*, **14**, 131 (1960).

8. N. Cabrera and N. F. Mott, "Theory of the Oxidation of Metals," *Rep. Progr. Phys.*, **12**, 163 (1948).

9. E. A. Benjamini, E. F. Duffek, C. A. Mylroie, and F. Schulenburg, "The Mobile Ionic Species During the Anodic Oxidation of Silicon in a Non-Aqueous Medium," Electrochemical Society October 1963 Meeting, New York.

10. L. Pauling, *The Nature of the Chemical Bond*, 3rd ed., Cornell Univ. Press, 1960, p. 85.

11. J. R. Ligenza, "Effect of Crystal Orientation on Oxidation Rates of Silicon in High Pressure Steam," *J. Phys. Chem.*, **65**, 2011 (1961).

12. P. J. Jorgensen, "Effect of an Electric Field on Silicon Oxidation," *J. Chem. Phys.*, **37**, 874 (1962).

PROBLEMS

2.1 Show that when a silicon dioxide film of thickness x_o is formed, a layer of silicon of thickness $0.45 x_o$ is consumed.

2.2 A silicon sample has a surface area of 1 cm².
 (a) Calculate the weight of a 0.2-μ thick oxide layer on this sample.
 (b) If the oxide layer contains 5×10^{16} excess oxygen molecules/cm³, calculate the total weight of the excess oxygen.

2.3 Calculate (a) the flux of oxidant, and (b) the time required to form a monolayer of SiO_2 in dry oxygen at 1200°C, when the oxide thickness is $0.2\ \mu$.

2.4 Using the general relationship, Equation 2.8, and the data given in the figures, construct the oxide thickness versus time curves for (a) dry oxygen at 1200°C; and (b) water vapor at 1000°C.

2.5 Using the curves obtained above, determine the oxide thickness:
 (a) After oxidation for one hour in dry oxygen at 1200°C.
 (b) After oxidation for 75 minutes in water vapor at 1000°C.
 (c) After step (a) + step (b).
 (d) After step (b) + step (a).

2.6 A silicon sample is covered with a 0.2-μ thick oxide layer. What is the additional time required to grow $0.1\ \mu$ more oxide in dry oxygen at 1200°C?

2.7 Calculate k_s for oxidation in dry oxygen and in water vapor at 1200°C (use the values of C^* given in the text). Compare these values with the surface-reaction rate constant in epitaxial growth.

2.8 Derive the relationship for thermal oxidation for the case when Henry's law at the oxide-silicon interface cannot be assumed. As an example, assume that the concentration of the oxidant at the outer surface of the oxide is proportional to the *square-root* of the partial pressure of the oxidant in the gas next to the oxide surface.
   (a) Derive a relationship between oxide thickness and time assuming diffusion of the oxidant across the oxide is infinitely rapid.
   (b) Repeat, assuming reaction at the oxide-silicon interface is infinitely rapid.
   (c) Derive the relationship in the general case. (This part involves considerable algebra.)

- **FLUX**
- **TRANSPORT EQUATION**
- **DIFFUSED LAYERS**
- **DEVIATIONS FROM SIMPLE DIFFUSION THEORY**
- **REDISTRIBUTION DURING OXIDATION**
- **DIFFUSION THROUGH SiO$_2$ LAYER**
- **REDISTRIBUTION IN EPITAXIAL GROWTH**

3
Solid-State Diffusion

The principal aim of semiconductor technology is to control the type and concentration of impurities within specific regions of a semiconductor crystal. The most practical way of achieving this is through *solid-state diffusion*. As a result, the diffusion of various impurities in semiconductors has been studied very extensively. Such studies have been facilitated by the fact that the electrical characteristics of *p-n* junctions and other semiconductor devices formed by solid-state diffusion can be directly employed in the measurement of the impurity concentrations. Using such electrical measurements, concentrations in the range of 10 parts per billion can be determined relatively accurately.

In this chapter, we discuss solid-state diffusion, with emphasis on the diffusion of impurities in silicon. We begin by deriving the formula giving the flux of a charged species, and the transport equation which determines their distribution in a solid. Then we discuss the method of formation and evaluation of diffused layers, using simple diffusion theory. Deviations from this theory due to such factors as the two-dimensional geometry of planar semiconductor devices, the charged nature of the impurities, and external rate limitations are then considered.

The second half of the chapter contains a treatment of other solid-state diffusion problems encountered in the fabrication of semiconductor devices. These include the redistribution of impurities during thermal oxidation, the diffusion of impurities through a silicon dioxide layer, and the redistribution of impurities during epitaxial growth.

3.1 FLUX

The flux F of any species is defined as $F \equiv$ number passing through unit area in unit time.

To derive a formula for the flux F, let us consider the example of the motion of positively charged impurities in a crystal. The atoms of the crystal form a series of potential hills which impede the motion of the charged impurities. This condition is represented in Figure 3.1a. The

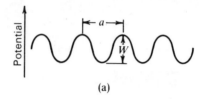

(a)

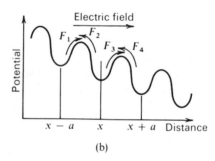

(b)

Fig. 3.1 Model of ionic motion within a crystal; potential distribution with and without applied bias.

height of the potential barrier W is typically of the order of electron volts in most materials. The distance between successive potential barriers a is of the order of the lattice spacing which is typically several angstroms.

If a constant electric field is applied, the potential distribution as a function of distance will be tilted, as shown in Figure 3.1b. This will make the passage of positively charged particles to the right easier and it will make their passage to the left more difficult. Let us now calculate the flux F at position x. This flux will be the average of the fluxes at position $(x - a/2)$ and at $(x + a/2)$. In turn, these two fluxes are given by $F_1 - F_2$ and $F_3 - F_4$, respectively, as indicated in Figure 3.1b.

Consider the component F_1. It will be given by the product of (i) the density per unit area† of impurities at the potential valley at $(x - a)$; (ii)

† This "area" is along a plane which is normal to the direction of the flux.

Flux

the probability of a jump of any of these impurities to the valley at x; and (iii) the frequency of attempted jumps ν. Thus we can write

$$F_1 = [aC(x - a)] \cdot \exp - \frac{q}{kT}[W - \tfrac{1}{2}a\mathscr{E}] \cdot \nu \tag{3.1}$$

where $[aC(x - a)]$ is the density per unit area of particles situated in the valley at $(x - a)$, and the exponential factor is the probability of a successful jump from the valley at $(x - a)$ to the valley at x. Note the lowering of the barrier due to the electric field $\mathscr{E}$.

Similar formulas can be written for F_2, F_3, and F_4. When these are combined to give a formula for the flux F at position x, with the concentrations $C(x \pm a)$ approximated by $C(x) \pm a(\partial C/\partial x)$, we obtain

$$F(x) = -[\nu a^2 e^{-qW/kT}]\frac{\partial C}{\partial x}\cosh\frac{qa\mathscr{E}}{2kT} + [2a\nu e^{-qW/kT}]\,C\sinh\frac{qa\mathscr{E}}{2kT}. \tag{3.2}$$

An extremely important limiting form of this equation is obtained for the case when the electric field is relatively small, i.e., $\mathscr{E} \ll kT/qa$. In this case we can expand the *cosh* and the *sinh* terms in the above equation. Noting that $\cosh(x) \doteq 1$ and $\sinh(x) \doteq x$ for $x \to 0$, this results in the limiting form of the flux equation for a positively charged species,

$$F(x) = -D\frac{\partial C}{\partial x} + \mu\mathscr{E}C \tag{3.3}$$

where
$$D \equiv \nu a^2 e^{-qW/kT} \tag{3.3a}$$

and
$$\mu \equiv \frac{\nu a^2 e^{-qW/kT}}{kT/q}. \tag{3.3b}$$

Note that the mobility μ and the diffusivity D are related by

$$D = \frac{kT}{q}\mu. \tag{3.4}$$

This is the well-known *Einstein's relationship*.

It is customary to identify the contribution to the flux which is proportional to the concentration gradient as the *diffusion* term, while the contribution which is proportional to the concentration itself is referred to as the *drift* term.

A similar derivation can be made for the motion of negatively charged species. Such a derivation leads to an equation similar to Equation 3.3 except that the sign of the drift term is negative. These expressions, as well as other important formulas developed in this chapter, are summarized in Table 3.3 at the end of this chapter.

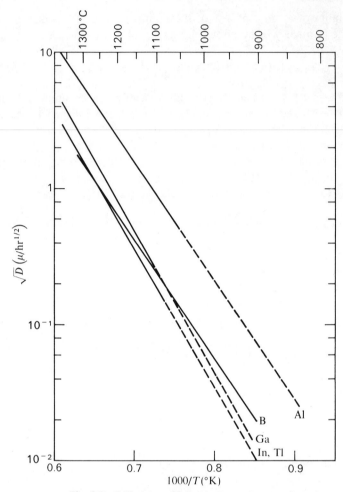

Fig. 3.2 Diffusivity of acceptor impurities in silicon.[1]

Experimentally measured values of the diffusivities of various impurities in silicon are shown in Figures 3.2 through 3.4. Diffusivities in silicon dioxide, in turn, are shown in Figure 3.5. Note that the logarithm of the diffusivity (or, equivalently, the logarithm of the square root of diffusivity which we show here for later convenience) plotted versus the reciprocal of the absolute temperature gives a good straight line in all cases. This implies that the temperature dependence of the diffusivity can be represented by the equation

$$D = D_o e^{-E_a/kT} \tag{3.5}$$

in agreement with Equation 3.3a.

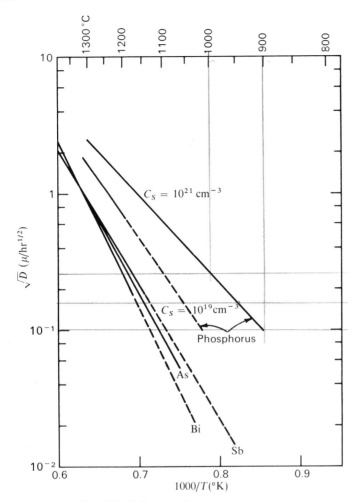

Fig. 3.3 Diffusivity of donor impurities in silicon.[1]

To assign a physical meaning to the activation energy E_a we must know the particular atomistic mechanism of diffusion. It is known that in semiconductors diffusion of *substitutional* impurities, i.e., impurities which occupy sites in the silicon lattice, usually proceeds by the impurities jumping into silicon vacancies in the lattice.[3] Thus the activation energy corresponds to the energy required to form a silicon vacancy rather than to the energy required to move the impurity. Since silicon-to-silicon bonds must be broken to form a vacancy, we might expect a relationship between the energy gap of a semiconductor (see Chapter 4) and the activation energy of the diffusivities of substitutional impurities. In fact, the activation energy for the diffusivities of acceptor and donor type impurities, which are all substitutional, all range between 3 and 4 ev in silicon, whereas the similar activation energies for diffusion in germanium all range between 2 and 3 ev—three to four times the energy gaps of Si and Ge, respectively.

Other impurities occupy the space between the silicon atoms, instead of the lattice sites in the crystal. Such *interstitial* impurities (e.g., manganese and nickel in silicon) generally move much more rapidly than substitutional impurities, as is evident from Figure 3.4. The atomistic mechanism of the

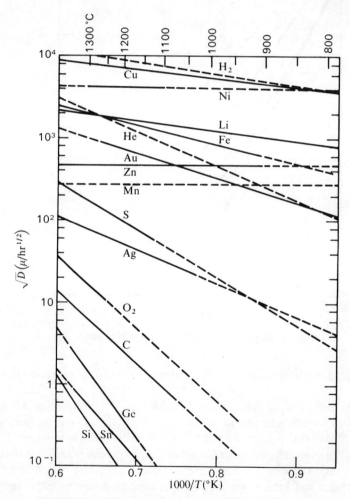

Fig. 3.4 Diffusivity of miscellaneous impurities in silicon.[1]

The Transport Equation

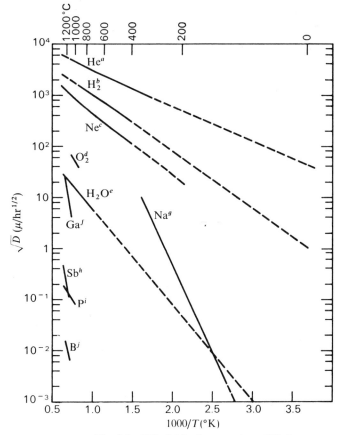

Fig. 3.5 Diffusivities in amorphous SiO_2.[2]

diffusion of gold in silicon is quite complex; it involves the motion of interstitial gold, but is influenced also by the concentration of vacancies.[4]

3.2 THE TRANSPORT EQUATION

Consider now a small element of a solid such as shown in Figure 3.6. An inventory of the material contained within the element bounded by the planes at position x and at position $x + \Delta x$ yields

$$\frac{\text{Increase in the density of material per unit area}}{\text{Unit time}} = F_{\text{in}} - F_{\text{out}},$$

provided material is neither formed nor consumed within this element.

In terms of concentrations, we can write

$$\Delta x \frac{\partial \bar{C}}{\partial t} = F(x) - F(x + \Delta x) \qquad (3.6)$$

where $\bar{C}$ is the average concentration within the element. Now, if we let $\Delta x \to 0$,

$$\bar{C} \to C(x), \quad \text{and} \quad \frac{F(x + \Delta x) - F(x)}{\Delta x} \to \frac{\partial F}{\partial x},$$

so that

$$\frac{\partial C}{\partial t} = -\frac{\partial F}{\partial x} \qquad (3.7)$$

which is the general form of the *transport equation* in one dimension.†

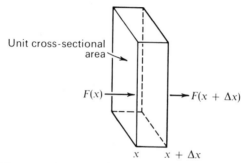

Fig. 3.6 The control element used in the derivation of the transport equation.

If we now substitute our formula for the flux of a positively charged species, Equation 3.3, and if we assume that the diffusivity is independent of the concentration, we arrive at a simpler form of the transport equation,

$$\frac{\partial C}{\partial t} = D \frac{\partial^2 C}{\partial x^2} - \mu \mathscr{E} \frac{\partial C}{\partial x} \qquad [\mathscr{E} \text{ and } D = \text{const.}]. \qquad (3.8)$$

In most of this chapter we concern ourselves with problems whose solutions we can approximate by assuming that the electric field $\mathscr{E} = 0$. This results in the *diffusion equation*,

$$\frac{\partial C}{\partial t} = D \frac{\partial^2 C}{\partial x^2} \qquad [\mathscr{E} = 0;\ D = \text{const.}]. \qquad (3.9)$$

† In three dimensions Equation 3.7 would become

$$\frac{\partial C}{\partial t} = -\text{div } \mathbf{F}.$$

Diffused Layers

The conduction of heat in solids is described by an identical equation except that the diffusivity D is replaced by the *thermal diffusivity*, κ. Thus we can use the solutions of heat conduction problems[5] in solid-state diffusion work.

3.3 DIFFUSED LAYERS

Solid-state diffusion is most frequently employed in semiconductor technology to form *diffused layers* of impurities. Typically, diffused layers are formed in a two-step process. In the first step, impurities are introduced into the semiconductor to a depth of a few tenths of a micron. This process is called the *predeposition* step. Once the impurities have been introduced into the semiconductor, they are then diffused deeper to provide a suitable concentration distribution without any more impurities being added to the semiconductor. This second step is called the *drive-in diffusion* step.

a. The Predeposition Step

The predeposition step is typically performed by placing the semiconductor sample in a furnace through which flows an inert gas containing the desired impurities.[6] The furnace and flow conditions are similar to those used in thermal oxidation (Chapter 2). The temperature usually ranges between 800° and 1200°C.

The impurities can be introduced into the carrier gas in several ways:[7]

Solid Source. In one arrangement the impurity is introduced into the carrier gas from a solid source which evaporates. If equilibrium is reached at the source, the partial pressure of the impurity compound in the gas will equal its vapor pressure at the source temperature.

A solid source can be used in the predeposition of most impurities. The source material is usually an oxide of the impurity, e.g., P_2O_5 for phosphorus, B_2O_3 for boron, As_2O_3 for arsenic, Sb_2O_4 for antimony. The oxide is carried to the semiconductor surface by the gas and then is reduced at the surface. An example of this reducing reaction might be

$$P_2O_5(\text{gas}) + Si(\text{solid}) \rightarrow P + SiO_2(\text{solid}).$$

Thus an oxide layer is formed on the silicon surface during predeposition.

Liquid Source. A liquid compound can also be used as a source of the impurities by passing the carrier gas through a flask containing the liquid. The flask and the liquid in it are kept at a fixed temperature. The gas which bubbles through the liquid reaches equilibrium with the liquid with the result that the partial pressure of the impurity compound in the gas leaving the flask will equal its vapor pressure at the temperature of the liquid. Liquid sources are available for several impurities, e.g., $POCl_3$ for phosphorus predeposition, and BBr_3 for boron predeposition.

Chemical Transport. A third method of introducing an impurity into the gas stream is through a chemical reaction taking place at the source. An example of this chemical transport process is that of gallium. The source employed in gallium predeposition can be gallium oxide, Ga_2O_3. It has been found[8] that gallium is introduced into the carrier gas only in the presence of hydrogen. This specificity indicates that a chemical reaction takes place at the source. One reasonable reaction might be

$$Ga_2O_3(\text{solid}) + 3H_2 \rightleftharpoons 2Ga(\text{vapor}) + 3H_2O.$$

For this reaction the partial pressure of gallium vapor should be given by

$$p_{Ga} \propto \left(\frac{p_{H_2}}{p_{H_2O}}\right)^{1.5}$$

according to the law of mass action. Experiments[9] show that the gallium concentration is, in fact, proportional to the 1.5-power of the hydrogen-to-water ratio, indicating that this reaction is the correct one.

If there is no resistance to the transport of the impurity from the gas to the solid surface, the surface concentration C_S in the solid will be in equilibrium with the partial pressure p of the impurity in the gas. Accordingly, if Henry's law is obeyed (see Chapter 2), we expect that

$$C_S = Hp.$$

There is, however, a maximum concentration of any impurity that can be accommodated in a solid at any given temperature. This concentration is called the *solid solubility* of the impurity. Thus the above relationship can hold only up to the limit of the solid solubility of the impurity in the semiconductor at the temperature of the predeposition. Solid solubilities of important impurities in silicon are shown in Figure 3.7 as a function of temperature.

The relationship between surface concentration and partial pressure is illustrated by Figure 3.8 which shows the measured surface concentration of boron in silicon as a function of the partial pressure of boron-oxide in the carrier gas for an 1100°C predeposition.[11] It is seen that the surface concentration is indeed proportional to the partial pressure in the gas at low concentrations, with an eventual saturation expected at the solid solubility limit.

Diffused Layers

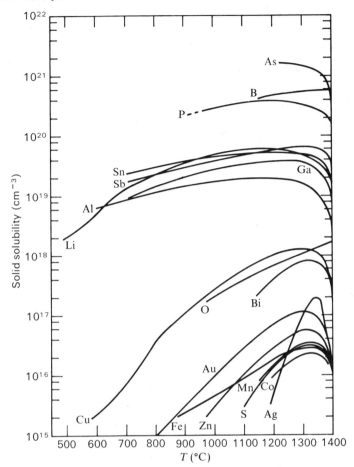

Fig. 3.7 Solid solubilities in silicon.[10]

In industrial practice, predeposition is usually performed with a partial pressure of the impurity in the carrier gas high enough so that the surface concentration in the semiconductor corresponds to the solid solubility of the impurity. Under such conditions the surface concentration is independent of the partial pressure of the impurity and will be reproducible and uniform even if the partial pressure is not.

We can now proceed to the consideration of the impurity distribution in the semiconductor resulting after predeposition. Since during predeposition the surface concentration C_S of the impurity is kept constant, we must solve the diffusion equation (3.9) subject to the boundary conditions

$$C(0, t) = C_S \tag{3.10}$$

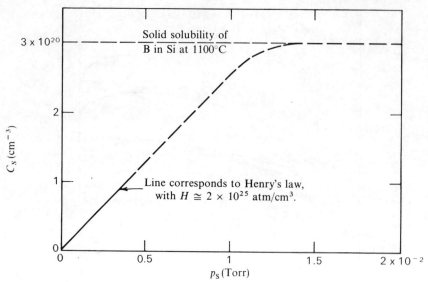

Fig. 3.8 Surface concentration of boron in silicon as a function of the partial pressure of B_2O_3 in the ambient, at 1100°C.[11]

and
$$C(\infty, t) = 0 \tag{3.11}$$
and the initial condition
$$C(x, 0) = 0. \tag{3.12}$$

The solution satisfying the diffusion equation as well as the above conditions is the *complementary error function*,[5]
$$C(x, t) = C_S \operatorname{erfc} \frac{x}{2\sqrt{Dt}} \tag{3.13}$$

which is one of the most important functions encountered in diffusion theory. Some of its properties are summarized in Table 3.1. It is tabulated in Carslaw and Jaeger[5] as well as in many mathematics handbooks.

The concentration distribution resulting in a predeposition process according to the above simple theory is shown in Figure 3.9 where we plot, on both linear and logarithmic scales, the normalized concentration (normalized to the surface concentration, which is constant) as a function of distance for three different values of the *diffusion length* $2\sqrt{Dt}$ corresponding to three consecutive predeposition times. Note that as the predeposition time progresses, the impurity penetrates deeper and deeper into the solid. This penetration can be represented by a useful quantity,

Diffused Layers

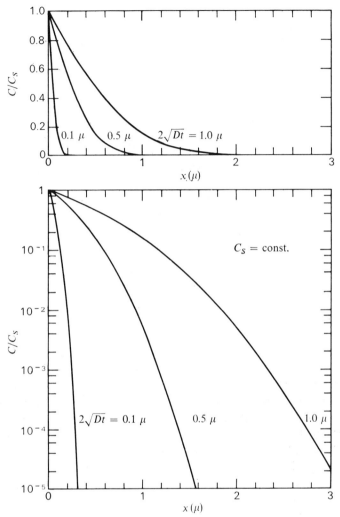

Fig. 3.9 The complementary error function (erfc): normalized concentration versus distance for successive times.

the total number of impurity atoms in a 1-cm² section of the semiconductor, defined by

$$Q(t) \equiv \int_0^\infty C(x, t)\, dx. \quad (3.14)$$

Integration of Equation 3.13 yields

$$Q(t) = \frac{2}{\sqrt{\pi}} \sqrt{Dt}\, C_S. \quad (3.15)$$

TABLE 3.1
SOME PROPERTIES OF THE ERROR FUNCTION

$$\operatorname{erf}(x) \equiv \frac{2}{\sqrt{\pi}} \int_0^x e^{-a^2} \, da$$

$$\operatorname{erfc}(x) \equiv 1 - \operatorname{erf}(x)$$

$$\operatorname{erf}(0) = 0$$

$$\operatorname{erf}(\infty) = 1$$

$$\operatorname{erf}(x) \cong \frac{2}{\sqrt{\pi}} x \quad \text{for } x \ll 1$$

$$\operatorname{erfc}(x) \cong \frac{1}{\sqrt{\pi}} \frac{e^{-x^2}}{x} \quad \text{for } x \gg 1$$

$$\frac{d \operatorname{erf}(x)}{dx} = \frac{2}{\sqrt{\pi}} e^{-x^2}$$

$$\int_0^x \operatorname{erfc}(x') \, dx' = x \operatorname{erfc} x + \frac{1}{\sqrt{\pi}}(1 - e^{-x^2})$$

$$\int_0^\infty \operatorname{erfc}(x) \, dx = \frac{1}{\sqrt{\pi}}$$

This expression can be interpreted simply as follows. The quantity $Q(t)$ represents the *area* under one of the concentration distribution curves of the top portion of Figure 3.9. These distributions can, in turn, be approximated by triangles whose height is C_S and whose base is $\sim 2\sqrt{Dt}$. This leads to $Q(t) \cong \sqrt{Dt} \, C_S$, which is reasonably close to the exact form, Equation 3.15.

Another quantity of considerable importance in determining the characteristics of semiconductor devices is the *gradient* of the impurity concentration $\partial C/\partial x$. It can be obtained by differentiating Equation 3.13. This leads to

$$\left.\frac{\partial C}{\partial x}\right|_{(x,t)} = -\frac{C_S}{\sqrt{\pi Dt}} e^{-x^2/4Dt}. \tag{3.16}$$

Using the asymptotic approximation for the complementary error function given in Table 3.1, which is valid under most practical conditions, this can be rewritten as

$$\left.\frac{\partial C}{\partial x}\right|_{(x,t)} \cong -\frac{x}{2Dt} C(x, t). \tag{3.17}$$

b. Drive-In Diffusion

As discussed above, predepositions usually result in surface concentrations which equal the solid solubility of the impurity. It is often necessary to lower the surface concentration from this value. Furthermore, it is often desirable to move the impurities deeper into the semiconductor without at the same time increasing the total number of impurities within the semiconductor.

Both purposes can be accomplished by a further high-temperature heat treatment in a gas which does not contain any impurities. In practice, this additional heat treatment, the drive-in diffusion, is usually carried out in an oxidizing ambient.

A comparison of the data shown in Figures 3.2 and 3.3 with those shown in Figure 3.5 indicates that most acceptor- and donor-type impurities diffuse much faster in silicon than in silicon dioxide. (Gallium is a notable exception.) Thus, if the heat treatment is performed in an oxidizing ambient, the resulting oxide layer will tend to seal these impurities into the silicon; no more impurities will be added nor will a significant amount of impurities escape because the oxide acts as an effective mask against the diffusion of these impurities.

The concentration distribution after drive-in diffusion is accordingly given by that solution of the diffusion equation which satisfies the boundary conditions

$$\left.\frac{\partial C}{\partial x}\right|_{(0,t)} = 0 \tag{3.18}$$

and

$$C(\infty, t) = 0 \tag{3.19}$$

which provide for a constant Q being maintained in the semiconductor during the drive-in diffusion step. The initial condition is given by

$$C(x, 0) = C_S \,\text{erfc}\, \frac{x}{2\sqrt{(Dt)_{\text{predep}}}} \tag{3.20}$$

since the impurity distribution at the beginning of the drive-in diffusion step is the impurity distribution resulting at the conclusion of the predeposition step.

This is a difficult problem to solve. However, for most practical cases, $\sqrt{Dt}$ for drive-in diffusion is much larger than $\sqrt{Dt}$ for predeposition. Thus we can regard the extent of penetration of the predeposited profile to be negligibly small in comparison to that of the final profile resulting after the drive-in diffusion step. Accordingly, we can represent the predeposition profile mathematically by a *delta function*.

The concentration distribution after drive-in diffusion subject to this approximation then will be given by[5]

$$C(x, t) = \frac{Q}{\sqrt{\pi Dt}} e^{-x^2/4Dt}. \tag{3.21}$$

This is the well-known *Gaussian* distribution. Note that in this solution the total amount of impurities in the solid, Q, is constant by virtue of the boundary condition Equation 3.18. Since the impurities move into the solid with the passage of time, in order to keep the total Q constant the concentration near the surface must drop. Indeed, it is evident from Equation 3.21 that the surface concentration is given by

$$C_S(t) = \frac{Q}{\sqrt{\pi Dt}}. \tag{3.22}$$

Thus the concentration distribution can also be represented by

$$C(x, t) = C_S(t) e^{-x^2/4Dt}. \tag{3.23}$$

The distribution of impurities for a Gaussian profile is shown in Figure 3.10 where we plot the concentration of the impurities normalized by Q, which is a constant, as a function of distance for three increasing diffusion lengths or three successive drive-in diffusion times. Note that the slope of each impurity distribution at $x = 0$ is zero corresponding to the first boundary condition, and that the diffusion profile spreads inward into the semiconductor, with the surface concentration dropping as the diffusion time increases.

The gradient of the impurity concentration is obtained by differentiating Equation 3.23 and is

$$\left.\frac{\partial C}{\partial x}\right|_{(x,t)} = -\frac{x}{2Dt} C(x, t). \tag{3.24}$$

It is important to note that for both the complementary error-function and the Gaussian distributions the concentration is a function of a normalized distance, $x/2\sqrt{Dt}$. Thus, if we normalize the concentration with the surface concentration, we can represent the distributions with a single curve each, valid for all times, as shown in Figure 3.11. It is important to realize that in the case of the Gaussian distribution we now have also introduced the time variable into our other normalizing parameter, the surface concentration.

Let us now consider the exact solution of this problem without the delta-function approximation to the initial distribution. Such a solution[12]

Diffused Layers

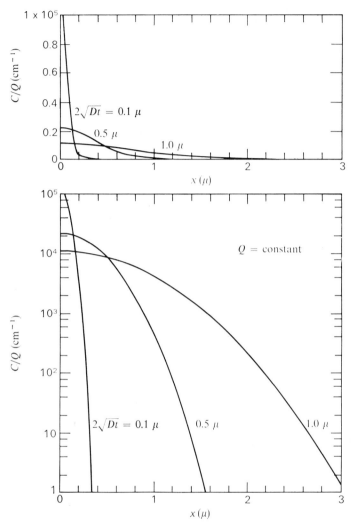

Fig. 3.10 The Gaussian: normalized concentration versus distance for successive times.

is shown in Figure 3.12, in comparison to the corresponding distribution obtained with the delta-function approximation. In this figure the concentration normalized by the constant surface concentration present during the predeposition step is plotted as a function of the distance normalized with the diffusion length of the drive-in diffusion step. The parameter is the ratio of the diffusion length of the predeposition step to that of the drive-in diffusion step. Note that for values of this ratio as large as 0.25, the delta-function approximation provides an excellent fit to the more

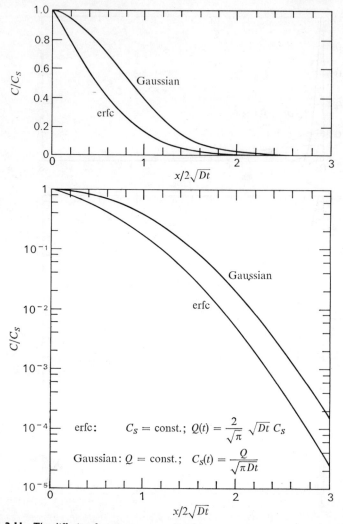

Fig. 3.11 The diffusion functions: normalized concentration versus normalized distance.

exact results. Only when this ratio becomes of the order of 1, does the delta-function approximation begin to lead to large errors.

c. Evaluation of Diffused Layers

Routine evaluation of diffused layers is relatively simple if the diffused layer forms a *p-n* junction with the underlying substrate. The depth of this junction x_j can be delineated by grooving into the semiconductor and

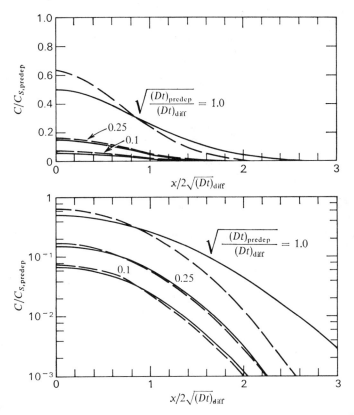

Fig. 3.12 Two-step diffusion: predeposition followed by drive-in diffusion. Solid lines are exact final distributions.[12] Dashed lines correspond to case when predeposited distribution is approximated by delta-function at $x = 0$.

etching the surface with a solution which reacts selectively with the two sides of the junction, thus resulting in a color variation between the p and the n sides as illustrated in Figure 3.13. This method makes possible the determination of the junction depth x_j with about a 0.1-micron accuracy using optical techniques.[13]

The junction depth x_j, as illustrated in Figure 3.13, is defined by the condition

$$C(x_j, t) = C_B. \tag{3.25}$$

Thus, if the junction depth and the bulk concentration C_B are known, the surface concentration C_S as well as the rest of the impurity distribution can be calculated, provided we can safely assume that the concentration distribution follows one or the other simple diffusion functions. An

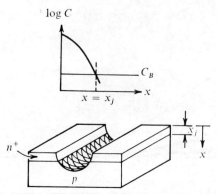

Fig. 3.13 Determination of junction depth by grooving and staining.

important and convenient check on the validity of this assumption is provided by a measurement of the *average resistivity* of the diffused layer. The reciprocal of the average resistivity $\bar{\rho}$ of an n-type layer (taken as example) is

$$\frac{1}{\bar{\rho}} = \frac{q}{x_j} \int_0^{x_j} \mu_n n(x)\, dx \tag{3.26}$$

where we are essentially adding up the parallel conductances of the elemental segments of the diffused layer as indicated in Figure 3.14.

To evaluate the integral appearing in this equation, we must have (i) a relationship between the mobility and the impurity concentration, and (ii) the carrier concentration as a function of distance. For complete ionization of impurities (see Chapter 4), this will be approximately the same as the distribution of the impurities.

We now illustrate the procedure of evaluating $\bar{\rho}$ by considering the simplest of all possible distributions: that corresponding to a "box" profile illustrated in the inset to Figure 3.15. In the case of a "box" distribution the integral in Equation 3.26 can be evaluated with ease and results in

$$\frac{1}{\bar{\rho}} = q\mu_n(C_S - C_B) \tag{3.27}$$

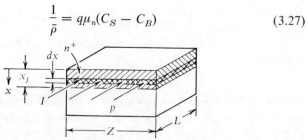

Fig. 3.14 Determination of the average resistivity of diffused layers.

Diffused Layers

which we can rearrange to yield

$$C_S = C_B + \frac{1}{q\mu_n \bar{\rho}}. \tag{3.28}$$

If we take $\mu_n = 85$ and $\mu_p = 45$ cm²/v sec which correspond to the limiting mobility values reached at high impurity concentrations in silicon, such as we are likely to encounter in the case of diffused layers (see Chapter 4),

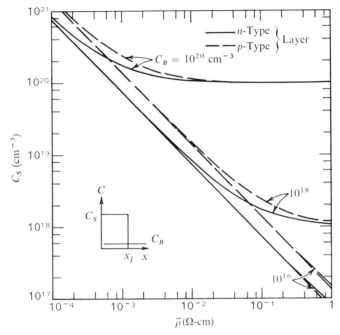

Fig. 3.15 Surface concentration versus average resistivity—"box" distribution.

we can obtain the relationships between surface concentration and average resistivity for a "box" distribution for both n- and p-type layers. These are shown in Figure 3.15.

The box distribution, of course, is not usually a realistic one. The integral in Equation 3.26 has also been evaluated numerically for the two diffusion functions, the complementary error function and the Gaussian, for the case of silicon.[14] The resulting curves, analogous to the one in Figure 3.15 for the "box" distribution, are shown in Figure 3.16, where we again show surface concentration C_S as a function of average resistivity $\bar{\rho}$, for different substrate concentrations C_B. Note that these curves, which are based on computer calculations and take the experimentally

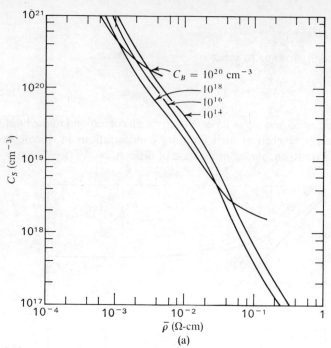

Fig. 3.16a Surface concentration versus average resistivity[14]—n-Type layer, erfc distribution.

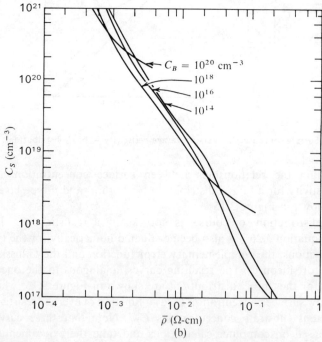

Fig. 3.16b Surface concentration versus average resistivity[14]—n-Type layer, Gaussian distribution.

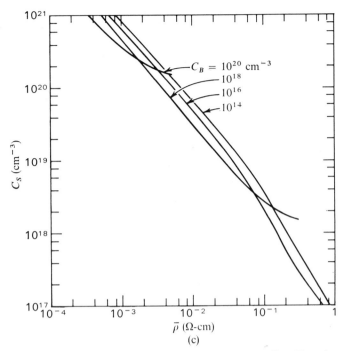

Fig. 3.16c Surface concentration versus average resistivity[14]—p-Type layer, erfc distribution.

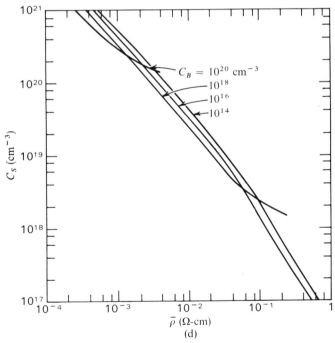

Fig. 3.16d Surface concentration versus average resistivity[14]—p-Type layer, Gaussian distribution.

measured variation of the electron and hole mobilities with impurity concentration into account (see Chapter 4), have the same general tendency as those obtained by the crude "box" profile approximation.

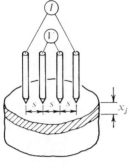

The average resistivity of diffused layers is customarily measured by the *four-point probe* technique[15] illustrated in Figure 3.17. In this technique four probes are placed on the surface of the semiconductor. Current is passed through the two outermost probes and the voltage—the IR drop—due to this current flow is measured between the two middle probes. Thus any problems due to probe-to-semiconductor contact resistances are eliminated. In order to relate the measurements of the voltage drop V and the current I to the average resistivity $\bar{\rho}$ of the diffused layer we must solve the electrostatic problem which takes into account the spreading of the current from the two outside probes as shown in the top view in Figure 3.17. This has been done and the result for layers whose lateral dimensions are large and whose width x_j is small in comparison to the spacing of the probes is[16]

Top view

Fig. 3.17 Measurement of average resistivity using a four-point probe.

$$\bar{\rho} = \frac{\pi}{\ln 2} \frac{V}{I} x_j = 4.532 \frac{V}{I} x_j. \qquad (3.29)$$

Since a typical probe spacing is $s = 1000\ \mu$, this condition is amply met for diffused layers.† If surface concentrations based on measurements of the average resistivity agree with those obtained from junction depth measurements, we can conclude that our diffusion profile does indeed follow the assumed distribution.

3.4 DEVIATIONS FROM SIMPLE DIFFUSION THEORY

We now examine some of the causes of deviation of the impurity distributions from the simple theory considered in Section 3.3.

† Because semiconductor wafers are generally sliced 200 to 400 μ thick, Equation 3.29 can also be used to calculate wafer resistivity from four-point probe measurements by substituting the wafer thickness for x_j. If either the wafer thickness or the lateral dimension of the semiconductor sample is comparable to the probe spacing, correction factors must be applied.[16]

a. Two-Dimensional Diffused Layers

In the planar technology, the diffusion process is usually carried out through windows cut in a diffusion mask (usually an SiO_2 layer). As a result, the impurities will diffuse *parallel* to the surface as well as normal to the surface. The resulting junction shape near the edge of the window is observed experimentally to be approximately cylindrical.

The results of detailed calculations[17] of this two-dimensional diffusion problem for the predeposition case are shown in Figure 3.18. Here we plot the normalized concentration distribution as a function of distance parallel to the surface from the edge of the opening, and also as a function of distance normal to the surface. The latter is identical with the complementary error-function type of distribution. The distribution parallel to the surface deviates from the error-function distribution but this deviation leads to no more than about 15 or 20% error in the location of a junction. Thus, within this error, the junction shape can indeed be regarded as cylindrical. The calculations for the case of a drive-in diffusion lead to similar conclusions.

b. Field-Aided Diffusion

When donor or acceptor impurities enter the silicon crystal, they become ionized. Consequently, we have to consider the simultaneous motion of the ionized donors (or acceptors) and electrons (or holes).

We have already discussed the simultaneous motion of two charged species in SiO_2 in connection with oxidation kinetics in Chapter 2. We concluded then that whenever two charged species which have different mobilities move simultaneously, a built-in electric field will result, aiding the motion of the slower of the two species.

Detailed consideration of the problem of the motion of donor or acceptor ions in a semiconductor[18] leads to a modified formula for the flux of the ions,

$$F = -D\left[1 + \frac{1}{\sqrt{1 + 4(n_i/C)^2}}\right]\frac{\partial C}{\partial x} \qquad (3.30)$$

where C is the concentration of the ions and n_i denotes the concentration of electrons or holes in a semiconductor containing no donors or acceptors at all, at the temperature of diffusion. (This quantity, called the *intrinsic carrier concentration*, is discussed in Chapter 4.)

Equation (3.30) can also be written in the form

$$F = -D_{\text{eff}}\frac{\partial C}{\partial x} \qquad (3.31)$$

where

$$D_{\text{eff}} \equiv D\left[1 + \frac{1}{\sqrt{1 + 4(n_i/C)^2}}\right]$$

is the effective diffusivity of the impurities, which incorporates the effect of the built-in electric field. Note that this effective diffusivity is a function of the concentration.

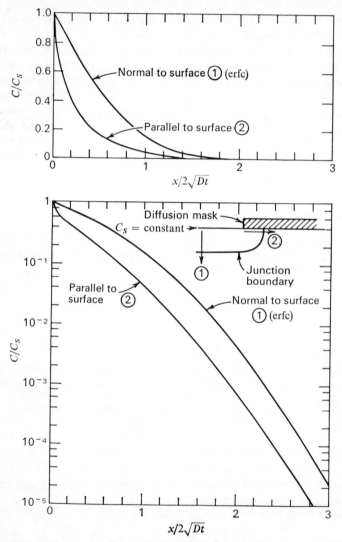

Fig. 3.18 Comparison of the concentration distributions ① normal, and ② parallel to the surface, for conditions corresponding to predeposition.[17]

We can now consider the two limiting values of the effective diffusivity D_{eff}. At a given temperature, hence at a given value of n_i, if the impurity concentration is relatively low, i.e., $C \ll n_i$, $D_{\text{eff}} = D$. Thus, the built-in field has no effect. In the opposite case, when the impurity concentration is high, i.e., $C \gg n_i$, $D_{\text{eff}} = 2D$. Thus, the electric field can bring about an effective doubling of the diffusion coefficient.

The effect of the built-in field on the shape of the concentration distribution is illustrated in Figure 3.19 where the results of numerical calculations[18] based on the equations for field-aided diffusion, are shown for a particular case. For comparison, we also show a complementary error-function distribution, which does not take the field-aiding effect into consideration and which is fitted to the correct distribution at low concentrations. It is apparent that the field-aiding effect brings about a smaller gradient near the surface. This is because the electric field aids the motion of the ions to the largest extent in the region of highest impurity concentration. Thus in this region the concentration gradient does not need to be so steep to maintain the same flux flowing into the semiconductor.

Experimentally measured[19] concentration distributions of boron in silicon under various surface-concentration conditions are shown in Figure 3.20. It is evident that for the lowest surface concentration the distribution closely follows the complementary error-function distribution. It also appears that the distribution begins to deviate from the error-function shape as the surface concentration exceeds 10^{19} cm^{-3}. (Note that n_i in silicon at 1100°C, the diffusion temperature, is approximately 10^{19} cm^{-3}.) The intermediate profile could be accounted for in terms of the field-aided diffusion effect as shown in Figure 3.19. However, as the surface concentration of boron is increased further to exceed 10^{20} cm^{-3}, the profile begins to deviate from the complementary error-function distribution to a much larger extent than could be accounted for in terms of the field-aiding effect. It is evident that we must consider other causes of deviation in such a case.

c. Effect of Lattice Strain

One important anomaly that is observed when the surface concentration exceeds a certain value is the commonly known "emitter-push" effect. (It is alternately called the "emitter-dip" effect or "cooperative diffusion" effect.) This phenomenon is illustrated in Figure 3.21. It consists of the enhanced diffusion of a diffused layer in regions where a second, high-concentration diffused layer penetrates the silicon. It is often observed in transistor structures which, by necessity, must employ two consecutive

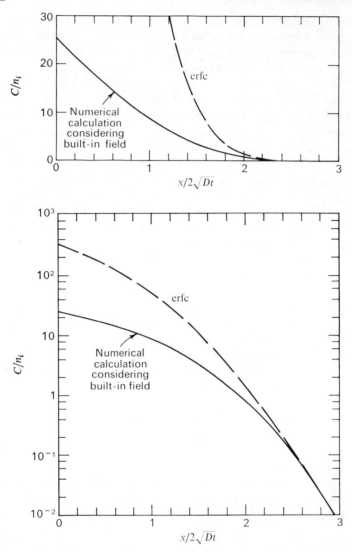

Fig. 3.19 Illustration of the effect of a built-in field on the concentration distribution in a semiconductor.[18]

diffusions with the second one, the emitter, having a high surface concentration (hence the name).

In Figure 3.21 the boron distribution in silicon is shown both under a high concentration phosphorus-diffused region and elsewhere in the

Deviations from Simple Diffusion Theory

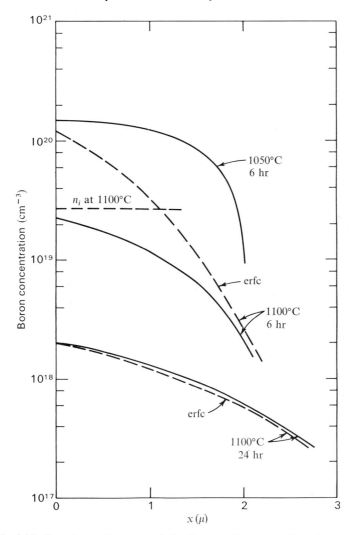

Fig. 3.20 Experimentally measured distribution of boron in silicon for various surface concentrations.[19]

sample. It is evident that the phosphorus diffusion "pushes" the boron distribution ahead of it. This brings about the irregular junction boundary illustrated in the inset.

Several possible explanations of this effect have been proposed.[20] It appears most likely that straining of the lattice by the high concentration of impurities is involved. This proposition was supported by an experiment[20]

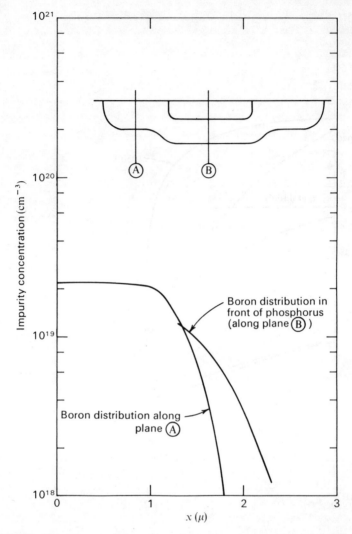

Fig. 3.21 The "emitter-push" effect: anomalous diffusion due to a second, high concentration, diffusion.[20]

in which a pressure probe was placed on the silicon surface during the diffusion process. It was found that under the pressure probe the impurities did indeed penetrate deeper into the silicon than elsewhere.

This effect leads to a very serious limitation in the fabrication of high-frequency transistors which, as we shall see later, require close spacing between two diffusion fronts.

d. External Rate Limitation of Solid-State Diffusion

In our treatment of the problem of predeposition, we have assumed that the surface concentration of the impurity C_S is a constant throughout the time of predeposition. Initially, however, there are no impurities in the semiconductor and therefore the surface concentration must also be zero. Because physically the surface concentration cannot jump instantaneously from zero to its final constant value, there will have to be a period of adjustment during which the simple boundary condition $C(0, t) = C_S$ will not hold. The duration of this initial period is determined by the relative rates of (i) transport of impurities from the bulk of the gas to the surface, and (ii) solid-state diffusion from the surface into the interior of the semiconductor. Thus an external limitation to the rate of transport of the impurities will influence the solid-state diffusion process.

We next consider the solution of the predeposition problem subject to such an external rate limitation. The initial condition subject to which we have solved the predeposition problem,

$$C(x, 0) = 0 \tag{3.32}$$

as well as one of the boundary conditions,

$$C(\infty, t) = 0 \tag{3.33}$$

remain unchanged. However, we now replace the second boundary condition, that of a constant surface concentration, with the condition

$$h[C_S - C(0, t)] = -D \frac{\partial C}{\partial x}\bigg|_{(0,t)} \tag{3.34}$$

which specifies that the flux to the surface via mass transfer in the gas will equal the flux away from the surface via solid-state diffusion. As in Chapter 2, h denotes the gas-phase mass-transfer coefficient in terms of concentrations in the solid.

The solution to this problem[21]

$$\frac{C(x, t)}{C_S} = \operatorname{erfc} \frac{x}{2\sqrt{Dt}} - e^{[ht/\sqrt{Dt}]^2} \cdot e^{(ht/\sqrt{Dt})(x/2\sqrt{Dt})} \operatorname{erfc} \left[\frac{ht}{\sqrt{Dt}} + \frac{x}{2\sqrt{Dt}} \right]. \tag{3.35}$$

This solution consists of a complementary error function from which a correction term is subtracted. As the parameter $ht/\sqrt{Dt}$ becomes very large, the subtracted term vanishes. Calculations based on Equation 3.35 are shown in Figure 3.22. It is evident that as $ht/\sqrt{Dt}$ approaches approximately 10, the solution merges into the complementary error function.

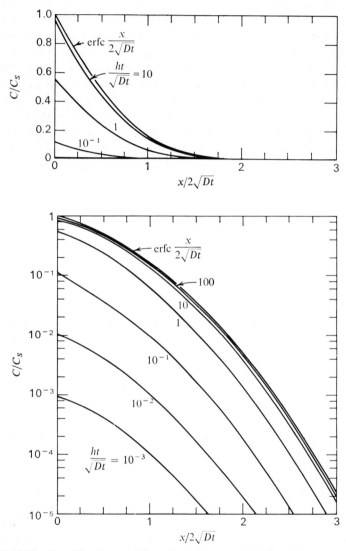

Fig. 3.22 The effect of external rate limitation on solid-state diffusion: predeposition of impurities.[21]

Thus the dimensionless parameter $ht/\sqrt{Dt}$ provides a simple estimate of the period of time during which this adjustment of the surface concentration takes place. This time will be of the order of $t \simeq 100(D/h^2)$. As an estimate, we can take $h_G = 5$ cm/sec for the gas-phase mass-transfer coefficient, from Chapter 1. The mass-transfer coefficient in terms of concentrations in the solid $h = h_G/HkT$. If we base the value of the Henry's law constant on the boron data shown in Figure 3.8, $H = 2 \times 10^{25}$ atm/cm^3, we get $h \simeq 2 \times 10^{-6}$ cm/sec, or about 70 μ/hr. For $D \simeq 1$ μ^2/hr, this leads to $t \simeq 0.02$ hr or a little over a minute. Since predepositions are often of only a few minutes duration, the influence of gas-phase rate limitation may become appreciable.

The complementary problem to that of predeposition is the escape of impurities from a semiconductor, often referred to as "out-diffusion." The initial condition for this problem will be a uniform distribution corresponding to the bulk concentration C_B,

$$C(x, 0) = C_B \tag{3.36}$$

and the boundary conditions will be

$$C(\infty, t) = C_B \tag{3.37}$$

and

$$hC(0, t) = -D \frac{\partial C}{\partial x}\bigg|_{(0,t)}. \tag{3.38}$$

Here we assume that the concentration of the impurities in the bulk of the gas is zero. The solution of this problem is very similar to the one for the case of predeposition. It is given by[21]

$$\frac{C(x, t)}{C_B} = \mathrm{erf}\frac{x}{2\sqrt{Dt}} + e^{[ht/\sqrt{Dt}]^2} e^{(ht/\sqrt{Dt})(x/2\sqrt{Dt})} \mathrm{erfc}\left[\frac{ht}{\sqrt{Dt}} + \frac{x}{2\sqrt{Dt}}\right]. \tag{3.39}$$

Note that this solution consists of a simple error function to which a term dependent upon the parameter $ht/\sqrt{Dt}$ is added. As this parameter becomes large, the additional term becomes very small. Calculations based on Equation 3.39 are shown in Figure 3.23. It is evident that the solution converges to the simple error function for $ht/\sqrt{Dt} > 10$.

Experimental measurements[22] of the impurity distribution in a silicon sample resulting after exposure to a hydrogen ambient are shown in Figure 3.24. It is evident that most of the experimental points lie below the theoretical curve corresponding to $ht/\sqrt{Dt} = 5$. For the time and temperature of these experiments, this indicates that the value of h in this case is larger than about 10 μ/hr, which is consistent with the above estimate of 70 μ/hr.

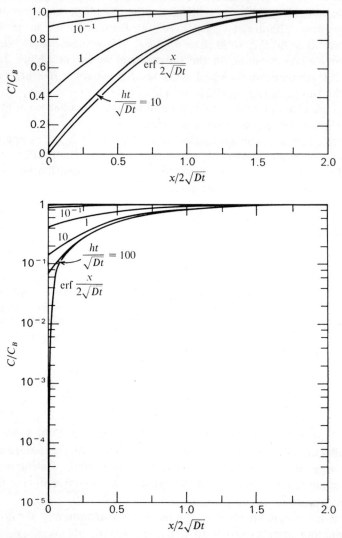

Fig. 3.23 The effect of external rate limitation on solid-state diffusion: escape (out-diffusion) of impurities.[21]

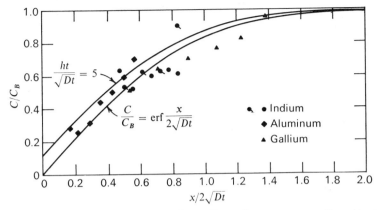

Fig. 3.24 Outdiffusion of impurities from silicon after exposure to H_2 ambient for 30 minutes at 1275°C.[22]

3.5 REDISTRIBUTION OF IMPURITIES DURING THERMAL OXIDATION

Whereas the previous considerations dealt with deviations from the boundary condition applied during the predeposition step, in this section we consider a phenomenon which can cause deviations from the boundary condition applied in our treatment of the drive-in diffusion step, $\partial C/\partial x = 0$ at $x = 0$.

To arrive at this boundary condition we have reasoned that, because most impurities diffuse much slower in silicon dioxide than in silicon, the oxide layer grown during the drive-in diffusion step will effectively seal the impurities into the silicon. In reality, the situation at the growing silicon dioxide-silicon interface is considerably more complex.

It has been found[23] that impurities in the silicon will be redistributed near a growing thermal oxide. This is due to several factors. If any two phases—solid, gas, or liquid—are brought into contact, an impurity contained in one of the two phases will be redistributed between the two until equilibrium is reached. In equilibrium, the ratio of the concentrations in the two phases will be a constant. The ratio of the equilibrium concentration in the silicon to that in the silicon dioxide is denoted by the term *segregation coefficient* and is defined as

$$m \equiv \frac{\text{Equilibrium concentration of impurity in silicon}}{\text{Equilibrium concentration of impurity in SiO}_2}.$$

Another factor that influences the process of impurity redistribution during thermal oxidation is that the impurity may have a tendency to escape through the oxide layer. If the diffusivity of the impurity in the oxide is small, this factor will be unimportant; however, if the diffusivity is relatively large, this factor can significantly affect the impurity distribution in the silicon.

A third factor entering into the redistribution process is the fact that the oxide layer is growing and therefore the boundary separating the oxide

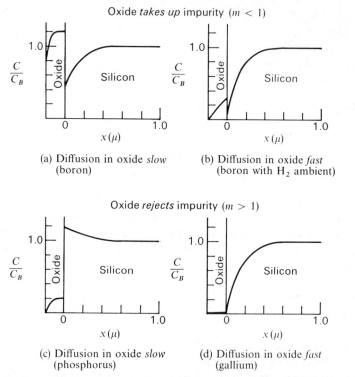

Fig. 3.25 Four different cases of impurity redistribution in silicon due to thermal oxidation.[24]

and silicon phases is moving as a function of time. The relative rate of this motion as compared to the diffusion rate is important in determining the extent of redistribution.

Note that even if the segregation coefficient of an impurity $m = 1$, some redistribution of the impurity in the silicon will still take place. As pointed out in Chapter 2, the oxide layer will fill up more space than the silicon used in the oxidation. Thus the same amount of impurity will now be distributed in a larger volume, bringing about a depletion of the impurity near the silicon surface.

Redistribution of Impurities during Thermal Oxidation

Four possible cases of the redistribution process[24] are illustrated in Figure 3.25. These cases can be considered in two groups: in one, the oxide has a tendency to *take up* impurity, and in the other the oxide has a tendency to *reject* the impurity. In each case the situation can be different for *slow* and for *fast* diffusion in the oxide. In Figure 3.25 an experimentally observed example is listed for each of the four cases.

We now consider the starting premise and the results of the theory of the redistribution process.[24] This theory is based on a solution of the diffusion equation with a moving boundary and describes the redistribution of an initially uniform doping concentration in the silicon. The boundary conditions applied to the problem are: (i) the concentration of the impurity at the gas-oxide interface is a constant, C_o; (ii) far within the silicon substrate the concentration approaches the bulk concentration C_B. In addition, we have to satisfy two *matching conditions*. The first is that the concentrations at the two sides of the oxide-silicon interface must be in the ratio prescribed by the segregation coefficient m, and the second is that, as the oxide grows, the impurity must be conserved at the moving oxide-silicon interface. It is also assumed that the oxide growth will proceed proportionally to the square root of the oxidation time, i.e., $x_o = \sqrt{Bt}$ where B is the parabolic oxidation rate constant (see Chapter 2).

The solution giving the impurity distributions in the silicon and in the silicon dioxide layer is in terms of error functions and complementary error functions. It is not reproduced here. The most important feature of this solution is the resulting formula for the concentration of the impurity on the silicon side of the interface,

$$\frac{C_S}{C_B} = \frac{1 + (C_o/C_B)\lambda}{1 + (1/m - \alpha)\sqrt{\pi} \exp(\alpha^2 B/4D) \operatorname{erfc}(\alpha\sqrt{B/4D})\sqrt{B/4D} + \lambda/m} \quad (3.40)$$

where

$$\lambda \equiv r \exp[(\alpha^2 r^2 - 1)B/4Dr^2] \operatorname{erfc}(\alpha\sqrt{B/4D})/\operatorname{erf}(\sqrt{B/4D_o})$$

and

$$r \equiv \sqrt{D_o/D}.$$

D_o, D are diffusivities of the impurity in oxide and silicon, respectively, and α is the ratio of the thickness of silicon consumed during oxidation to the oxide thickness. Its value is 0.45.

One interesting feature of the theory of redistribution is that, as evident from Equation 3.40, the concentrations at the moving oxide-silicon interface will be *independent of oxidation time*. In simple terms, both oxidation and diffusion proceed with $\sqrt{\text{time}}$ and, because of this, the time variable disappears from the expression for surface concentration. Thus a steady state is reached at the interface.

Equation 3.40 shows that, in line with our earlier qualitative discussion, the surface concentration in the silicon is a function of (i) the segregation coefficient m, (ii) the relative rates of diffusion in the silicon and in the oxide D_o/D, and (iii) the relative rates of oxidation to diffusion, B/D.

The role of the first two of these factors has already been illustrated in Figure 3.25. The third one is illustrated in Figure 3.26 for phosphorus and in Figure 3.27 for boron where the surface concentrations of these

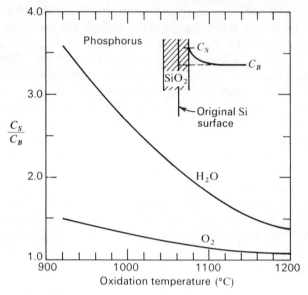

Fig. 3.26 Surface concentration of phosphorus in silicon after thermal oxidation. Calculated using Equation 3.40, and $m = 10$.

impurities, calculated by using Equation 3.40 and the experimentally determined values of the segregation coefficient m, are shown as a function of oxidation temperature, for oxidation in O_2 and H_2O ambients. It is apparent that as the speed of oxidation is increased—for example, by oxidizing in water vapor rather than in oxygen at a given temperature—the extent of redistribution will be increased as evidenced by the higher surface concentration of phosphorus and by the lower surface concentration of boron.

These two figures give the surface concentration as a function of oxidation temperature. The *amount* of impurity lost from the silicon as a result of redistribution is another matter. Whereas the surface concentration depends on oxidation temperature as shown in the above two figures, the amount of impurity lost will depend also on the *extent* of the disturbance of the originally uniform impurity distribution. This, in turn, is given approximately by the diffusion length of the impurity, $2\sqrt{Dt}$. Figure 3.28 shows the calculated boron distribution profiles in silicon

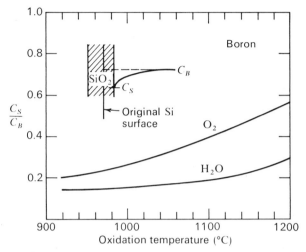

Fig. 3.27 Surface concentration of boron in silicon after thermal oxidation. Calculated using Equation 3.40, and $m = 0.3$.

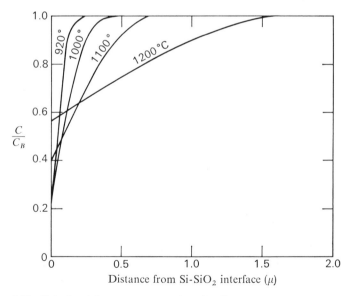

Fig. 3.28 Calculated boron concentration distribution in silicon after oxidation at various temperatures (O_2, $x_o = 0.2\ \mu$).[25]

after oxidation at various temperatures, with each oxidation resulting in an oxide layer 0.2 μ thick.[25] Note that as the oxidation temperature is decreased, the surface concentration of boron decreases in accordance with Figure 3.27. However, because the diffusivity of boron also decreases, the extent of the depression in concentration will also be smaller at the lower temperatures. These calculations were fully verified by experimental measurements.[25]

Table 3.2 shows the segregation coefficients of important acceptor and donor impurities as determined by several experimental techniques. For comparison, we also list values of these segregation coefficients which were predicted on the basis of solubility thermodynamic considerations.[26]

TABLE 3.2

SEGREGATION COEFFICIENTS OF IMPURITIES AT Si/SiO$_2$ INTERFACE

Impurity	m		
	Thermodynamic estimate[26]	Based on redistribution experiments	Based on oxide masking experiments[29]
Ga	>10^3		~20
B	10^{-3}–10^3	~0.3[24,25]	~10^{-2}
		0.1[27]	
In	>10^3		
P, Sb, As	>10^3	~10 [24,25]	

So far we have considered only the relatively simple problem of the redistribution of an originally uniform impurity concentration present in the silicon. An even more important practical problem is the redistribution of a predeposited layer during the drive-in diffusion step. Because the initial distribution is more complicated, this problem is more difficult to study both experimentally and theoretically. The results of a study[27] of the redistribution of diffused boron layers after thermal oxidation is shown in Figure 3.29. Here the theoretical line represents the results of numerical computations for an assumed segregation coefficient $m = 0.1$ while the experimental points were obtained by sectioning and successive measurements of the average resistivity of the boron-diffused layer. The solution is normalized to the surface concentration during the predeposition step which preceded the drive-in diffusion. It is evident that the results shown in this figure are in general agreement with what we would expect from the studies of the redistribution of a uniformly doped silicon substrate.

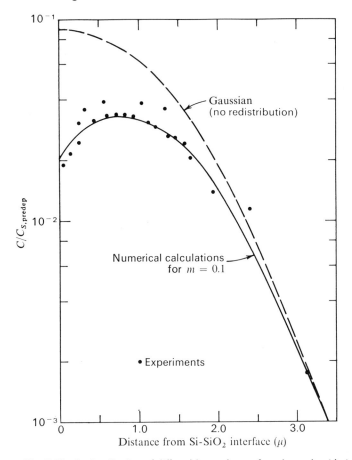

Fig. 3.29 Redistribution of diffused boron layer after thermal oxidation.[27]

3.6 DIFFUSION THROUGH A SILICON DIOXIDE LAYER (OXIDE MASKING)

The fact that the silicon dioxide layer is relatively impermeable to several acceptor and donor impurities[6] has been an exceedingly important factor in the development of the planar technology. In order to get a more quantitative picture of the masking phenomenon, we now consider the predeposition of an impurity through a silicon dioxide layer. We assume that the thickness of this layer remains constant throughout the predeposition. The expected distribution of the impurities after predeposition is illustrated schematically in Figure 3.30.

The initial condition for this diffusion problem is that the concentration of the impurity is zero both in the oxide and in the silicon. The boundary conditions are (i) that the concentration at the outer surface of the oxide is a constant C_o throughout the predeposition step (corresponding to the constant C_S in the case of predeposition onto a bare semiconductor sample considered earlier); and (ii) that the impurity concentration vanishes deep inside the semiconductor. In addition, we must satisfy the two matching

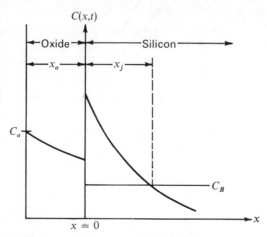

Fig. 3.30 Schematic illustration of the impurity distribution resulting after predeposition through an oxide layer.

conditions: (i) that the ratio of concentrations at the oxide-silicon interface is determined by the segregation coefficient m; and (ii) that the flux of impurities through this interface is continuous.

The solution of this problem is in the form of an infinite series.[28] However, for relatively small values of x such that $x \ll \sqrt{D/D_o}\, x_o$ we can approximate the infinite series by its first term alone. This results in

$$\frac{C(x, t)}{C_o} \simeq \frac{2mr}{m + r} \operatorname{erfc}\left[\frac{x_o}{2\sqrt{D_o t}} + \frac{x}{2\sqrt{Dt}}\right] \qquad (3.41)$$

where $r \equiv \sqrt{D_o/D}$.

A particularly important question is the depth of a junction formed in the silicon under a protective oxide layer when the impurity has diffused through this layer. The position of this junction is defined by the condition

Diffusion through a Silicon Dioxide Layer

which, when substituted into Equation 3.41, results in

$$\frac{x_j}{\sqrt{t}} = -\frac{1}{r}\frac{x_o}{\sqrt{t}} + I \tag{3.42}$$

where

$$I \equiv 2\sqrt{D}\, \text{arg}\left[\text{erfc}\,\frac{(m+r)C_B}{2mrC_o}\right].$$

Computer calculations[29] of the junction depth x_j as a function of the oxide thickness x_0, based on the infinite series, are shown in Figure 3.31 for

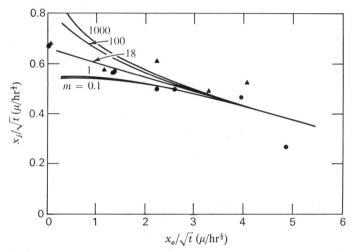

Fig. 3.31 Comparison between theoretical calculations and experimental results for gallium predeposition at 1100°C.[29]

the specific case of the predeposition of gallium at 1100°C. The curves were calculated for various assumed values of the segregation coefficient m. Note that regardless of the value of m all curves approach the straight-line relationship, Equation 3.42, for small values of x_j. The slope of this straight-line asymptote directly yields r. The *shape* of the experimental x_j versus x_0 relationship at relatively large values of x_j in comparison with the family of theoretical curves enables the estimation of the segregation coefficient m. Segregation coefficients estimated in this manner from the experimental data on gallium[29] and boron[30] predeposition through an oxide film are included in Table 3.2.

A technologically important quantity, the minimum oxide thickness required to prevent the formation of a junction in the silicon, the *masking thickness*, is readily calculated from Equation 3.42 by setting $x_j = 0$.

3.7 THE REDISTRIBUTION OF IMPURITIES IN EPITAXIAL GROWTH

In epitaxial growth the film contains either different impurities than the substrate, or the same impurities but in different concentrations. It is usually desirable that the doping concentration gradients between film and substrate be sharp. However, epitaxial growth has to be performed at

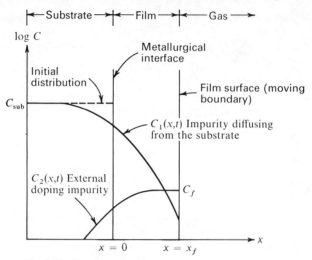

Fig. 3.32 Impurity distribution in epitaxial growth.

elevated temperatures (see Chapter 1) and therefore the diffusion of impurities will be relatively rapid. Since diffusion will tend to even out the concentration gradients at the interface between epitaxial film and substrate, the diffusion of impurities during epitaxial growth is of obvious practical importance.

The impurity distribution near the epitaxial film-substrate interface consists of two components as illustrated in Figure 3.32. One, whose concentration is designated by C_1, is due to the impurity diffusing out from the substrate; the other, whose concentration is designated by C_2, is the external doping impurity which is intentionally incorporated into the growing film and which, during growth, diffuses inward into the substrate. The total distribution of impurities is determined by the *sum* of these two components if they are impurities of the same type and by their *difference* if they are impurities of opposite types. In the latter case a *p-n* junction will be formed at the intersection of the two impurity concentration distributions.

The Redistribution of Impurities in Epitaxial Growth

The solid-state diffusion problem corresponding to the substrate impurity is solved subject to the initial condition

$$C_1(x, 0) = C_{\text{sub}} \tag{3.43}$$

and the boundary condition,

$$C_1(-\infty, t) = C_{\text{sub}} \tag{3.44}$$

which designate that the initial distribution in the substrate is a constant C_{sub} corresponding to the substrate doping concentration and that the impurity distribution deep inside the substrate remains undisturbed as the epitaxial growth proceeds.

The second boundary condition is given by

$$-D\frac{\partial C_1}{\partial x} = (h + V)C_1 \quad \text{at } x = x_f = Vt, \tag{3.45}$$

where h is the gas-phase mass-transfer coefficient in terms of concentrations in the solid, and V is the film-growth rate. Equation 3.45 is a statement of the conservation of impurities at the advancing film surface. Because of its importance we will derive this boundary condition carefully.

The total amount of substrate impurities per unit area contained within the semiconductor substrate and film is given by

$$Q(t) \equiv \int_{-\infty}^{x_f} C_1(x, t)\, dx. \tag{3.46}$$

The time rate of decrease of $Q(t)$ is, in turn, given by

$$-\frac{dQ}{dt} = -\int_{-\infty}^{x_f} \frac{\partial C_1}{\partial t} dx - VC_1(x_f, t) = hC_1(x_f, t) \tag{3.47}$$

where we employed Leibnitz's rule of differentiating integrals. The last portion of the equation is the expression for the escape of impurities from a solid surface to a gaseous ambient, if the concentration of the impurity in the gas is zero.

Next, we note that because of the diffusion equation we have

$$\int_{-\infty}^{x_f} \frac{\partial C_1}{\partial t} dx = D\int_{-\infty}^{x_f} \frac{\partial^2 C_1}{\partial x^2} dx = D\left[\frac{\partial C_1}{\partial x}\bigg|_{x=x_f} - \overset{0}{\overbrace{\frac{\partial C_1}{\partial x}\bigg|_{x=-\infty}}}\right]. \tag{3.48}$$

Thus, by combining Equation 3.48 with 3.47, we arrive at our boundary condition, Equation 3.45.

The solution[22] of this problem is fairly complicated, especially since this last boundary condition is specified at a moving boundary. For our purposes, this solution can be represented as

$$\frac{C_1(x, t)}{C_{\text{sub}}} = f_1\left[\frac{x}{2\sqrt{Dt}}, \frac{ht}{\sqrt{Dt}}, \frac{Vt}{\sqrt{Dt}}\right] \quad (3.49)$$

which shows that the normalized concentration distribution is a function of the dimensionless distance, an escape-rate parameter, and a growth-rate parameter. For $V = 0$, the solution reduces to the externally rate-limited outdiffusion case which we discussed earlier in this chapter.

More importantly, however, the solution has a very simple limiting form for the case of $Vt/\sqrt{Dt} \gg 1$. In that case the solution is given to a very good approximation by

$$\frac{C_1(x, t)}{C_{\text{sub}}} \doteq \tfrac{1}{2} \operatorname{erfc} \frac{x}{2\sqrt{Dt}}, \quad (3.50)$$

regardless of the value of h.

Calculations based on the complete solution, for the case $ht/\sqrt{Dt} \gg 1$ and for various values of the parameter $Vt/\sqrt{Dt}$, are shown in Figure 3.33. It is evident that the impurity distribution approaches the simple form given by Equation 3.50 for values of $Vt/\sqrt{Dt}$ larger than about 5. Since typical epitaxial film thicknesses are of the order of about 10 μ, whereas typical values of $\sqrt{Dt}$ might be of the order of less than 1 μ, typical values of the growth parameter $Vt/\sqrt{Dt}$ would be larger than 10—thus "infinitely" large for the purposes of these calculations. Hence, the simple solution given by Equation 3.50 provides a very good approximation for practical epitaxial growth conditions. (The same conclusion was reached on the basis of calculations performed for values of the escape rate parameter $ht/\sqrt{Dt}$ between 10^{-3} and 10^3.)

It is easy to give an interpretation to this simple limit of the exact solution. The film growth rate is so high relative to the rate of diffusion from the substrate that, insofar as the diffusion profile is concerned, the film grows to an "infinite" thickness almost instantaneously. Thus the concentration distribution will be very close to that which is obtained in the problem of diffusion between two semi-infinite slabs[5]—Equation 3.50.

The diffusion of the external doping impurity is described by a solution of the diffusion equation subject to the initial condition

$$C_2(x, 0) = 0 \quad (3.51)$$

The Redistribution of Impurities in Epitaxial Growth

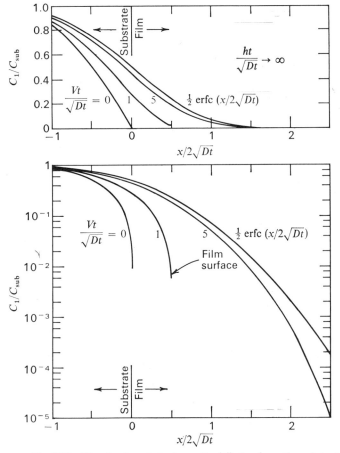

Fig. 3.33 Distribution of the impurity diffusing from the substrate.[22]

and the boundary conditions

$$C_2(-\infty, t) = 0 \tag{3.52}$$

$$C_2(x_f, t) = C_f. \tag{3.53}$$

The first boundary condition states that deep inside the substrate the concentration of the external doping impurity vanishes. The second specifies that the concentration of the external doping impurity at the growing film surface is a constant C_f (C_f is determined by the concentration of the impurity in the gas mixture). The solution to this problem[22] is also fairly complicated. It can be represented as

$$\frac{C_2(x, t)}{C_f} = f_2\left[\frac{x}{2\sqrt{Dt}}, \frac{Vt}{\sqrt{Dt}}\right]. \tag{3.54}$$

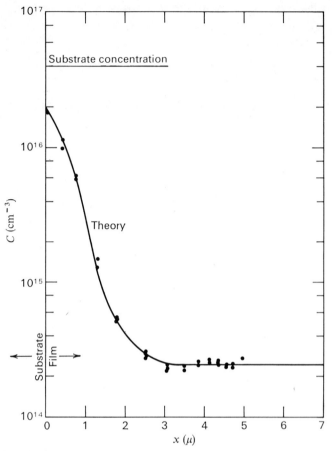

Fig. 3.34 Impurity distribution after epitaxial growth (antimony, 20 min. at 1275°C)[22].

For $Vt/\sqrt{Dt} \gg 1$ and for distances $x \gg 2\sqrt{Dt}$ this solution can be approximated simply by setting

$$C_2(x, t) \doteq C_f. \tag{3.55}$$

The combined solution describing the net distribution of impurities in epitaxial growth can then be obtained from

$$C(x, t) = C_1(x, t) \pm C_2(x, t) \tag{3.56}$$

where the positive sign is taken for impurities of the same type and the negative sign for impurities of opposite type.

Extensive experimental measurements[22] have verified the solution given above. An example of these results is shown in Figure 3.34.

In the above treatment we assumed that the impurity distribution in the epitaxial film is determined by the process of solid-state diffusion alone. This is indeed the case provided precautions are taken to eliminate contamination both from the reactor and from the back-side of the heavily doped substrates.[31] Impurities from the back-side of the substrates have been shown to be carried by the gas stream and incorporated into the growing epitaxial film on the front,[32] resulting in a much more gradual distribution of the substrate impurity than solid-state diffusion would. This phenomenon is often referred to as "autodoping."

READING REFERENCES

General treatments of solid-state diffusion are given by P. G. Shewmon, *Diffusion in Solids*, McGraw-Hill Book Co., 1963; R. M. Barrer, *Diffusion In and Through Solids*, Cambridge University Press, 1941; and W. Jost, *Diffusion in Solids, Liquids, Gases*, Academic Press, 1952.

A treatise on *Diffusion in Semiconductors* is given by B. I. Boltaks, Academic Press, 1963.

A review of diffusion data specifically dealing with silicon and of diffusion technology is found in Volume 4 of a series of review-reports on "Integrated Silicon Device Technology," by the Research Triangle Institute, ASD-TDR-63-316 (1964).

Diffusion data in glasses is reviewed by R. H. Doremus, "Diffusion in Non-crystalline Silicates," Chapter 1 in Volume 2 of *Modern Aspects of the Vitreous State*, J. D. Mackenzie, Ed., Butterworths, 1962.

An extensive compilation of the solutions of the diffusion equation under a wide variety of conditions can be found in H. S. Carslaw and J. C. Jaeger, *Conduction of Heat in Solids*, Oxford University Press, 2nd ed., 1959.

REFERENCES CITED

1. These diffusivities are based on Volume 4 of a series of review-reports on "Integrated Silicon Device Technology," by the Research Triangle Institute, ASD-TDR-63-316 (1964).
2. These diffusivities are based on:
    a. D. E. Swets, R. W. Lee, and R. C. Frank, "Diffusion Coefficients of Helium in Fused Quartz," *J. Chem. Phys.*, **34**, 17 (1961).
    b. R. W. Lee, R. C. Frank, and D. E. Swets, "Diffusion of Hydrogen and Deuterium in Fused Quartz," *J. Chem. Phys.*, **36**, 1062 (1962).
    c. R. C. Frank, D. E. Swets, and R. W. Lee, "Diffusion of Neon Isotopes in Fused Quartz," *J. Chem. Phys.*, **35**, 1451 (1961).
    d. F. J. Norton, "Permeation of Gaseous Oxygen through Vitreous Silica," *Nature*, **171**, 701 (1961).
    e. T. Drury and J. P. Roberts, "Diffusion in Silica Glass Following Reaction with Tritiated Water Vapor," *Phys. Chem. Glasses*, **4**, 79 (1963).

f. A. S. Grove, O. Leistiko, and C. T. Sah, "Diffusion of Gallium through a Silicon Dioxide Layer," *J. Phys. Chem. Solids*, **25**, 985 (1964).

g. A. E. Owen and R. W. Douglas, "The Electrical Properties of Vitreous Silica," *J. Soc. Glass Tech.*, **43**, 159 (1959).

h. M. O. Thurston and J. C. Tsai, "Diffusion of Impurities into Silicon through an Oxide Layer," Signal Corps Contract, Final Report DA-36-039-SC 87426 (1961).

i. C. T. Sah, H. Sello, and D. A. Tremere, "Diffusion of Phosphorus in Silicon Oxide Film," *J. Phys. Chem. Solids*, **11**, 288 (1959).

j. S. Horiuchi and J. Yamaguchi, "Diffusion of Boron in Silicon through Oxide Layer," *Jap. J. Appl. Phys.* **1**, 314 (1962).

3. B. I. Boltaks, *Diffusion in Semiconductors*, Academic Press, 1963.

4. W. R. Wilcox and T. J. LaChapelle, "Mechanism of Gold Diffusion into Silicon," *J. Appl. Phys.*, **35**, 240 (1964).

5. H. S. Carslaw and J. C. Jaeger, *Conduction of Heat in Solids*, Oxford University Press, 2nd ed., (1959).

6. C. J. Frosch and L. Derick, "Surface Protection and Selective Masking During Diffusion in Silicon," *J. Electrochem. Soc.*, **104**, 547 (1957).

7. An extensive review of silicon diffusion practices is given in Reference 1.

8. C. J. Frosch and L. Derick, "Diffusion Control in Silicon by Carrier Gas Composition," *J. Electrochem. Soc.*, **105**, 695 (1958).

9. G. E. Moore, unpublished.

10. F. A. Trumbore, "Solid Solubilities of Impurity Elements in Germanium and Silicon," *Bell System Tech. J.*, **39**, 205 (1960).

11. T. H. Yeh and W. Armstrong, "Diffusion of Boron in Silicon," Electrochemical Society Spring 1961 Meeting, Abstract No. 69, Indianapolis.

12. D. P. Kennedy and P. C. Murley, "Impurity Atom Distribution from a Two-Step Diffusion Process," *Proc. IEEE* (Corresp.), **54**, 620 (1964).

13. W. L. Bond and F. M. Smits, "The Use of an Interference Microscope for Measurement of Extremely Thin Surface Layers," *Bell System Tech. J.*, **35**, 1209 (1956).

14. J. C. Irvin, "Resistivity of Bulk Silicon and of Diffused Layers in Silicon," *Bell System Tech. J.*, **41**, 387 (1962).

15. L. B. Valdes, "Resistivity Measurements on Germanium for Transistors," *Proc. IRE*, **42**, 420 (1954).

16. F. M. Smits, "Measurement of Sheet Resistivities with the Four-Point Probe," *Bell System Tech. J.*, **37**, 711 (1958).

17. D. P. Kennedy and R. R. O'Brien, "Analysis of the Impurity Atom Distribution Near the Diffusion Mask for a Planar *P-N* Junction," *IBM Journal*, **9**, 179 (1965).

18. K. Lehovec and A. Slobodskoy, "Diffusion of Charged Particles into a Semiconductor under Consideration of the Built-in Field," *Solid-State Electron.*, **3**, 45 (1961).

19. S. Maekawa and T. Oshida, "Diffusion of Boron into Silicon," *J. Phys. Soc. Japan*, **19**, 253 (1964).

20. K. H. Nicholas, "Studies of Anomalous Diffusion of Impurities in Silicon," *Solid-State Electron.*, **9,** 35 (1966).
21. F. M. Smits and R. C. Miller, "Rate Limitation at the Surface for Impurity Diffusion in Semiconductors," *Phys. Rev.*, **104,** 1242 (1956).
22. A. S. Grove, A. Roder, and C. T. Sah, "Impurity Distribution in Epitaxial Growth," *J. Appl. Phys.*, **36,** 802 (1965).
23. M. M. Atalla and E. Tannenbaum, "Impurity Redistribution and Junction Formation in Silicon by Thermal Oxidation," *Bell System Tech. J.*, **39,** 933 (1960).
24. A. S. Grove, O. Leistiko, and C. T. Sah, "Redistribution of Acceptor and Donor Impurities During Thermal Oxidation of Silicon," *J. Appl. Phys.*, **35,** 2695 (1964).
25. B. E. Deal, A. S. Grove, E. H. Snow, and C. T. Sah, "Observation of Impurity Redistribution During Thermal Oxidation of Silicon Using the MOS Structure," *J. Electrochem. Soc.*, **112,** 308 (1965).
26. C. D. Thurmond, "Distribution Coefficients of Impurities Distributed Between Ge or Si Crystals and Ternary Alloys or Surface Oxides," in *Properties of Elemental and Compound Semiconductors*, H. C. Gatos, Ed., p. 121, Interscience, 1960.
27. T. Kato and Y. Nishi, "Redistribution of Diffused Boron in Silicon by Thermal Oxidation," *Jap. J. Appl. Phys.*, **3,** 377 (1964).
28. C. T. Sah, H. Sello, and D. A. Tremere, "Diffusion of Phosphorus in Silicon Oxide Film," *J. Phys. Chem. Solids*, **11,** 288 (1959).
29. A. S. Grove, O. Leistiko, and C. T. Sah, "Diffusion of Gallium Through a Silicon Dioxide Layer," *J. Phys. Chem. Solids*, **25,** 985 (1964).
30. S. Horiuchi and J. Yamaguchi, "Diffusion of Boron in Silicon Through Oxide Layer," *Jap. J. Appl. Phys.*, **1,** 314 (1962).
31. H. Basseches, S. K. Tung, R. C. Manz, and C. O. Thomas, "Factors Affecting the Resistivity of Epitaxial Silicon Layers," *Metallurgy of Semiconductor Materials*, **15,** 69 (1962).
32. B. A. Joyce, J. C. Weaver, and D. J. Maule, "Impurity Redistribution Processes in Epitaxial Silicon Wafers," *J. Electrochem. Soc.*, **112,** 1100 (1965).

PROBLEMS

3.1 Derive the transport equation in a manner similar to that employed in Section 3.2 for the case when the transported species is being consumed by a chemical reaction. Assume that the rate of this reaction at any point is proportional to the concentration of the species.

3.2 Verify that Equations 3.13 and 3.21 satisfy the diffusion equation and the appropriate initial and boundary conditions.

3.3 A phosphorus predeposition into silicon containing 10^{16} acceptor atoms/cm^3 resulted in an average resistivity of 4×10^{-4} Ω cm. Evaluate the phosphorus surface concentration, assuming
   (a) erfc
   (b) Gaussian   distribution.
   (c) "box"
Which is most realistic?

3.4 A p^+n junction diode was fabricated as follows:
Starting material: n-type, 5×10^{15} phosphorus/cm^3.
Predeposition: BCl_3, 990°C, 15 minutes.
Drive-in diffusion: Dry O_2, 1200°C, 9 hours.
  Measured $V/I = 60 \, \Omega$
  Measured $x_j = 7.0 \, \mu$
Calculate the impurity concentration as a function of distance, the total amount of diffused impurities per unit area (Q), and the impurity concentration gradient at the junction.

3.5 A commercial diode is fabricated as follows:
Starting material: n-type, 4×10^{14} phosphorus/cm^3.
Predeposition: BBr_3, 1150°C, 6 minutes.
Drive-in diffusion: Dry O_2, 1280°C, 16 hours.
  Measured $V/I = 0.8 \, \Omega$
  Measured $x_j = 37 \, \mu$
Evaluate as in Problem 3.4.
Compare the two diodes.

3.6 Derive Equation 3.30. In this derivation, assume that (i) the electric current is zero, (ii) space-charge neutrality prevails, and (iii) the product of electron and hole concentrations $pn = n_i^2$.

3.7 Using the intermediate results of the above problem, *estimate* the built-in electric field near the surface associated with the predeposition of boron at 1200°C, after 1 hour.

3.8 The different redistribution tendencies of impurities during thermal oxidation can be employed in the fabrication of useful device structures. As an example, oxidation of silicon initially containing both gallium and phosphorus in concentrations of 2×10^{16} cm^{-3} and 1×10^{16} cm^{-3}, respectively, will lead to the formation of an n-type skin under the oxide. If the oxidation is performed at 1200°C in dry oxygen, for 2 hours, calculate:
(a) The resulting impurity distributions.
(b) The depth of the junction, and the impurity concentration gradient at the junction.
(c) The average resistivity of the n-type skin.

3.9 Estimate the thickness of the silicon dioxide layer required to mask against the predeposition of boron at 1100°C, for a predeposition time of 1 hour. Assume that the solid solubility of boron in silicon dioxide is $(1/m)$ times its solid solubility in silicon. (Justify this assumption.)

3.10 An epitaxial film doped to contain 10^{16} boron atoms/cm^3 is grown upon four different substrates simultaneously. These substrates contain:
(a) 10^{14} phosphorus atoms/cm^3
(b) 10^{16} phosphorus atoms/cm^3
(c) 10^{18} phosphorus atoms/cm^3
(d) 10^{16} antimony atoms/cm^3.
Growth conditions are 20 minutes at 1250°C.
Calculate the junction depth for each wafer.

Problems

3.11 The base region of a *pnp* transistor is fabricated as follows:

    Starting material: *p*-type, 10^{19} boron/cm^3
    Epitaxial growth: 20 min. at 1250°C
                           Film growth rate: 1 μ/min.
                           Film doping: 5×10^{15} boron/cm^3
    Oxidation: 80 min. at 1200°C, water vapor.
Base window opened.
    Predeposition: P$_2$O$_5$, 30 min. at 800°C.
    Drive-in diffusion: 50 min. at 1200°C, dry oxygen
        Measured $V/I = 1.4 \, \Omega$
        Measured $x_j = 4 \, \mu$
Calculate:

(a) The oxide thickness over the base window and elsewhere.
(b) The impurity distribution after drive-in diffusion, including the distribution of the substrate impurity.
(c) What is the total number of impurities in the base (Q)? The gradient at the collector-base *n-p* junction? The distance over which the acceptor impurity concentration is uniform?

TABLE 3.3
IMPORTANT FORMULAS IN DIFFUSION

Flux:	
+ charge	$F = -D\dfrac{\partial C}{\partial x} + \mu \mathscr{E} C$
− charge	$F = -D\dfrac{\partial C}{\partial x} - \mu \mathscr{E} C$
Einstein's relationship:	$D = \dfrac{kT}{q}\mu$
Transport equation:	$\dfrac{\partial C}{\partial t} = -\dfrac{\partial F}{\partial x}$
for $\mathscr{E}$ = constant, + charge	$\dfrac{\partial C}{\partial t} = D\dfrac{\partial^2 C}{\partial x^2} - \mu \mathscr{E}\dfrac{\partial C}{\partial x}$
− charge	$\dfrac{\partial C}{\partial t} = D\dfrac{\partial^2 C}{\partial x^2} + \mu \mathscr{E}\dfrac{\partial C}{\partial x}$
Diffused layers	
constant C_S	$C(x, t) = C_S \operatorname{erfc} \dfrac{x}{2\sqrt{Dt}}$
	$Q(t) = \dfrac{2}{\sqrt{\pi}} \sqrt{Dt}\, C_S$
	$\left.\dfrac{dC}{dx}\right|_{(x,t)} = -\dfrac{1}{\sqrt{\pi Dt}} C_S e^{-x^2/4Dt} \cong -\dfrac{x}{2Dt} C(x, t)$
constant Q	$C(x, t) = \dfrac{Q}{\sqrt{\pi Dt}} e^{-x^2/4Dt}$
	$C_S(t) = \dfrac{Q}{\sqrt{\pi Dt}}$
	$\left.\dfrac{dC}{dx}\right|_{(x,t)} = -\dfrac{x}{2Dt} C(x, t)$

PART II

SEMICONDUCTORS AND SEMICONDUCTOR DEVICES

- **Elements of Semiconductor Physics**
- **Semiconductors under Non-Equilibrium Conditions**
- ***p-n* Junctions**
- **Junction Transistors**
- **Junction Field-Effect Transistors**

- **THE BAND THEORY OF SOLIDS**
- **ELECTRONS AND HOLES IN SEMICONDUCTORS**
- **FERMI-DIRAC DISTRIBUTION FUNCTION**
- **IMPORTANT FORMULAS**
- **TRANSPORT OF ELECTRONS AND HOLES**

4
Elements of Semiconductor Physics

In this chapter we present and discuss some important results of the physics of semiconductors. We consider only those aspects of semiconductor physics that are essential for the treatment of semiconductor devices and surface phenomena.

We begin with a qualitative discussion of the band theory of solids and of intrinsic and extrinsic semiconductors. Then we discuss the Fermi-Dirac distribution function and the concept of Fermi level, and present those important formulas of semiconductor physics which we will use throughout this book. Finally, we discuss the transport—drift and diffusion—of electrons and holes in semiconductors.

4.1 THE BAND THEORY OF SOLIDS

The most important result of the application of quantum mechanics to the description of electrons in a solid is that the allowed energy levels of electrons will be grouped into *bands*. The bands are separated by regions which designate energies that the electrons in the solid cannot possess. These regions are called *forbidden gaps*. The energy bands and a forbidden gap are illustrated schematically in Figure 4.1. The electrons in the outermost shell of the atoms comprising the solid, the *valence electrons*, are shown here in their lowest energy states. The band of these states is called the *valence band*.

The phenomenon of *conduction* is of principal interest in the study of semiconductor physics. Conduction consists of the motion of electrons. Thus conduction is possible only if we can get electrons into motion. In terms of energy considerations, this means that conduction is possible only if we can impart kinetic energy to an electron. We can now examine three different classes of solids—metals, insulators, and semiconductors—in terms of both an atomistic representation and the energy-band representation from the viewpoint of whether or not it is possible to energize an electron.

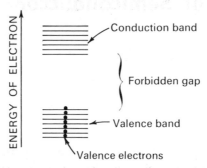

Fig. 4.1 Schematic energy band representation for electrons in a solid illustrating the energy bands and a forbidden gap.

In metals the valence electrons are free and constitute a sea of electrons which are free to move upon the application of even a small electric field. The corresponding energy-band representation with aluminum taken as an example is shown in the lower portion of Figure 4.2a. The two bands corresponding to the previous figure here overlap. Thus there is no forbidden gap. As a consequence, it is possible to move the topmost electron shown in this figure to the next level and then to the next and then to the next; in other words, it is possible to impart a kinetic energy to the electrons. Hence, conduction is possible.

If we now turn to the example of an insulator such as silicon dioxide, shown in Figure 4.2b, we are faced with a different situation. The valence electrons here form strong bonds between neighboring atoms. These bonds are difficult to break and, because of this, there will be no free electrons that could participate in conduction. In terms of the energy-band picture, this means that there is a large forbidden gap between the valence band and the next higher band, the *conduction band*. All levels in the valence band are occupied by electrons, all levels in the conduction

The Band Theory of Solids

band are empty. Because small electric fields cannot raise the uppermost electron in the valence band to the conduction band, it will be impossible to impart energy to any of the electrons shown in the band picture. For this reason, silicon dioxide will be an insulator—it will not conduct electricity.

The intermediate case of a semiconductor, with silicon taken as an example, is shown in Figure 4.2c. The bonds between neighboring silicon

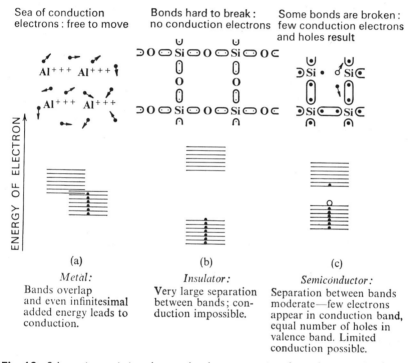

Fig. 4.2 Schematic atomistic and energy band representation of a conductor, an insulator, and an intrinsic semiconductor.

atoms are only moderately strong. Therefore, due to the thermal vibration of the silicon atoms, some bonds will be broken at any temperature above absolute zero. When a bond is broken, a free electron results which is capable of conducting electricity. Such an electron is called a *conduction electron*. In addition, there is now a "deficit" where the electron had been before the bond was broken. This deficit is referred to as a *hole*. Valence electrons can jump from neighboring bonds into the position of the hole and therefore additional conduction can take place. It is easy to think of this additional conduction as the movement of the positively charged hole in the opposite direction.†

In terms of the band representation, the forbidden gap of a semiconductor is not as large as that of an insulator. Because of this, some electrons will be able to make the jump from the valence band to the conduction band, leaving behind holes in the valence band. Upon the application of an electric field, the electrons in the conduction band will be able to gain kinetic energy; hence, they will be able to conduct electricity. At the same time the holes in the valence band will also be able to take on kinetic energy and conduct electricity.

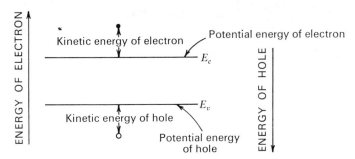

Fig. 4.3 Illustration of the meaning of the energy band diagram.

The energy-band diagram, such as shown in Figures 4.1 and 4.2, indicates electron energies. Thus when the energy of an electron is increased, the electron will take on a *higher* position in the band diagram. When we say that the energy of a hole is increased, what we mean is that the energy of the other electrons in the valence band is increased. Thus some of the valence electrons will take on higher positions in the band diagram. Accordingly, the increase in the energy of a hole is represented by the hole moving *downward* in the valence band. (Recall the "bubble analogy".)

It is important to note at this point that the lowest level in the conduction band designates the energy of a conduction electron which is at rest. The energy of an electron at rest is, of course, the potential energy of an electron, so the lower edge of the conduction band E_c designates the potential energy of an electron. Similarly, the upper edge of the valence band E_v designates the potential energy of a hole. If an electron is at a higher energy than the level E_c or a hole at a lower electron energy than the level

† The concept of a hole is analogous to that of a bubble in a liquid: although it is actually the liquid that moves, it is much easier to talk of the motion of the bubble in the opposite direction.

E_v, these electrons and holes have kinetic energies designated by the difference between their energies and the respective band edges, as illustrated in Figure 4.3.

4.2 ELECTRONS AND HOLES IN SEMICONDUCTORS

In absolutely pure semiconductors, conduction electrons and holes result only through the breakage of bonds. Thus the concentration of electrons n equals the concentration of holes p. These concentrations are called the *intrinsic carrier concentrations* n_i of the semiconductor. We would expect the intrinsic carrier concentration to be a function of the vibrational energy of the lattice (hence the temperature) that is responsible for the breakage of bonds. The intrinsic carrier concentration should also be a function of the energy required to break a bond, which in terms of the band diagram corresponds to the width of the forbidden gap, or energy gap, E_G.

Both of these dependences are borne out by the experimental data shown in Figure 4.4. Here the intrinsic carrier concentrations of three important semiconductors—gallium arsenide, silicon, and germanium—are shown as a function of temperature. The energy gaps of these semiconductors are also indicated in the figure. It is evident (i) that for any given semiconductor, n_i increases very sharply with increasing temperature; and (ii) that at any given temperature, n_i decreases very sharply with increasing energy gap. More detailed examination of these data indicates that both dependences can be summarized by the exponential temperature dependence, $n_i \propto e^{-E_a/kT}$ where the activation energy E_a is approximately $E_G/2$. This important relationship will be discussed further at a later stage.

Let us now consider the case where an impurity is incorporated into the single crystal semiconductor in concentrations which are much larger than n_i. In particular, let us first consider the case that arises when we add an impurity which has five valence electrons, such as phosphorus, to silicon which has four. This case is illustrated in Figure 4.5a. (Phosphorus occupies a place in the fifth column of the periodic table of the elements, while silicon occupies a place in the fourth column, in accordance with the respective number of their valence electrons.) The extra electron of the phosphorus atom cannot be accommodated in the regular bonding arrangement of the silicon lattice and, since it is out of place, it is easier to tear off. Thus the *ionization energy* of phosphorus in silicon is much smaller than the silicon energy gap. In fact, this ionization energy is only about 0.05 ev.

96 Elements of Semiconductor Physics

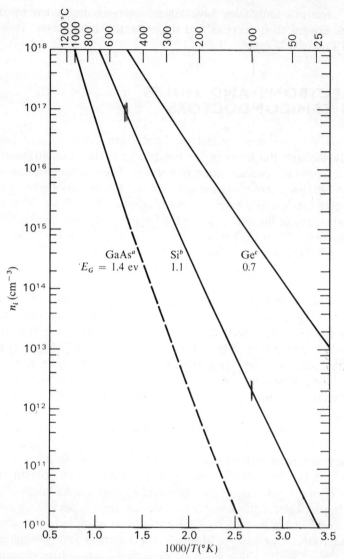

Fig. 4.4 Intrinsic carrier concentration of gallium arsenide, silicon, and germanium as a function of temperature.[1]

In a silicon crystal at room temperature there is usually enough lattice vibrational energy available to supply this amount of energy and, therefore, column V impurities in silicon will generally be all ionized at room temperature, providing an equal number of conduction electrons, unless they are present in relatively high ($>10^{18}$ cm^{-3}) concentrations. This condition is called *complete ionization*. Thus, under the condition of complete ionization, we can write that the concentration of electrons $n \doteq N_D$ where N_D denotes the concentration of the *donor* impurities. (Column V impurities in silicon are called donor impurities because they donate an electron to the conduction band of the silicon crystal.) This again is illustrated in the energy-band representation of Figure 4.5a where equal concentrations of electrons and donor ions are indicated. The donor ions are denoted by positive charges slightly below the conduction band edge in energy.

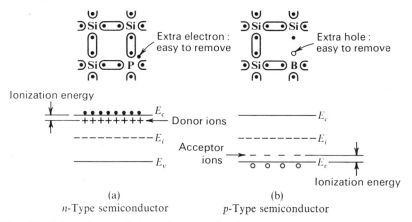

Fig. 4.5 Schematic atomistic and energy band representation of extrinsic semiconductors.

An analogous argument can be made for the case when an impurity which has three valence electrons (a column III impurity), such as boron, is introduced into the silicon lattice. Because a column III impurity has one less electron than silicon, we can consider it to carry a hole. This hole can then be removed relatively easily, with an ionization energy of approximately 0.05 ev.

If the ionization is complete, the concentration of the holes $p \doteq N_A$ where N_A denotes the concentration of *acceptor* impurities. (Column III impurities in silicon are called acceptor impurities because they can accept an electron from the valence band of the crystal. This, of course, is equivalent to supplying a hole to the valence band.) This situation is illustrated in Figure 4.5b, both in the atomistic and in the energy-band representations. Since we are considering the case of complete ionization, the concentration of holes in the valence band is shown to be equal to the concentration of acceptor ions in the crystal. The acceptor ions are denoted by negative charges slightly above the valence band edge in energy.

In the first of the above two cases, the concentration of electrons is much larger than that of holes. Because current in such a case is carried predominantly by electrons, we say that the conductivity type of the semiconductor is *n* (for negative). In the second case, the concentration of holes is much larger than the concentration of electrons and current is predominantly carried by holes. The conductivity type in such a case is *p* (for positive).

In general, both acceptor- and donor-type impurities may be present simultaneously. The conductivity type of the semiconductor is then determined by that impurity which is present in the greater concentration. The concentration of the corresponding *majority carrier* will then be given by $n \doteq N_D - N_A$ if $N_D > N_A$; and by $p \doteq N_A - N_D$ if $N_A > N_D$.

4.3 FERMI-DIRAC DISTRIBUTION FUNCTION

In the previous sections we have discussed various electronic energy states in semiconductors: states in the conduction and valence bands, and states introduced into the forbidden gap by the addition of donor or acceptor impurities. We now discuss what determines the probability that a given energy state is occupied by an electron.

The energy distribution of electrons in a solid are governed by the laws of *Fermi-Dirac statistics*. The principal result of these statistics is the *Fermi-Dirac distribution function* which gives the probability that an electronic state with energy E is occupied by an electron,

$$f(E) = \frac{1}{1 + e^{(E-E_F)/kT}}. \tag{4.1}$$

This function contains a parameter, E_F, which is called the *Fermi level*. A rigorous definition of the Fermi level describes it as the chemical potential of electrons in a solid. However, for our purposes it is sufficient to note that the Fermi level is that energy at which *the probability of occupation of an energy state by an electron is exactly one-half*.

The Fermi-Dirac distribution function is illustrated in Figure 4.6a for the case of an intrinsic semiconductor. At the left side of this figure we show the probability of occupation of states by electrons as a function of the energy of the states. In the conduction band there are a large number of states. However, the probability of occupation of these states is small; hence, there will be only a few electrons in the conduction band. By contrast, there are also a large number of states in the valence band. Most of these are occupied by electrons since the probability of occupation of states there is nearly unity. Thus there will be only few unoccupied electron states, i.e., holes, in the valence band.

Fermi-Dirac Distribution Function

The Fermi-Dirac distribution function is symmetrical around the Fermi level E_F. Thus, if the number of energy states in the conduction and valence bands is the same, and if the number of electrons in the conduction band and the number of holes in the valence band is also the same, the Fermi level must be located in the middle of the energy gap. This is approximately what happens in an intrinsic semiconductor. The Fermi level in an intrinsic semiconductor is often referred to as the *intrinsic Fermi level* and is denoted by the symbol E_i.

In an *n*-type semiconductor the concentration of electrons in the conduction band is larger than in the intrinsic case. Since, however, the density

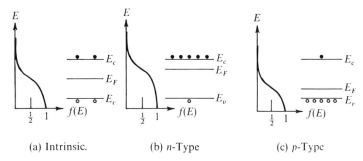

(a) Intrinsic. (b) *n*-Type (c) *p*-Type

Fig. 4.6 Illustration of the Fermi-Dirac distribution function for intrinsic, *n*- and *p*-type semiconductor.

of energy states in the conduction band is the same as in the intrinsic case, it follows that in an *n*-type semiconductor the Fermi level, and with it the entire Fermi-Dirac distribution function, will be shifted upward in the energy-band picture. In contrast, in a *p*-type semiconductor the Fermi level and the Fermi-Dirac distribution function will both be shifted downward. These two cases are illustrated in Fig. 4.6b and c.

For energies that are at least several kT units above or below the Fermi level, the Fermi-Dirac distribution function can be approximated by the simpler formulas

$$f(E) \doteq e^{-(E-E_F)/kT} \qquad \text{for } E > E_F \qquad (4.2)$$

and

$$f(E) \doteq 1 - e^{-(E_F-E)/kT} \qquad \text{for } E < E_F. \qquad (4.3)$$

It is useful to regard the second term of the last expression as the probability of occupation of a center located at energy E *by a hole*.

If, instead of Fermi-Dirac statistics, Boltzmann statistics had been employed in the derivation of the distribution function, these expressions would have resulted directly. Thus Boltzmann statistics provide a good approximation for Fermi-Dirac statistics for energies at least several kT away from E_F.

4.4 IMPORTANT FORMULAS FOR SEMICONDUCTORS IN EQUILIBRIUM

a. Electron and Hole Concentrations

If the densities of states in the conduction and valence bands are calculated on the basis of quantum mechanics, and if the probability of occupation of these states by electrons is taken into account, the electron and hole concentrations in a semiconductor can be calculated. Such a calculation yields the concentration of electrons,

$$n = N_c e^{-(E_c - E_F)/kT} \tag{4.4}$$

and the concentration of holes

$$p = N_v e^{-(E_F - E_v)/kT}. \tag{4.5}$$

We can recognize the exponential factor in the first of these expressions as the probability of occupation by an electron of a state located at the conduction-band edge E_c. The exponential factor in the second expression is the probability of occupation by a hole of a state located at the valence-band edge E_v. Accordingly, we can assign a simple interpretation to the pre-exponential factors: they designate the *effective densities of states* in the conduction and valence bands, respectively. Both N_c and N_v are proportional to $T^{3/2}$.

The values of the effective densities of states for germanium, silicon, and gallium arsenide at room temperature (27°C) are given in Table 4.1, along with other important properties of these semiconductors and of an insulator, silicon dioxide.

Equations 4.4 and 4.5 show, in accordance with the discussion in the previous section, that as the Fermi level E_F moves close to the conduction-band edge E_c, the concentration of electrons increases and the concentration of holes decreases. In contrast, when the Fermi level moves closer to the valence-band edge E_v, the electron concentration decreases and the hole concentration increases.

As the Fermi level approaches either the conduction- or the valence-band edge within a few kT units, the approximations to the Fermi-Dirac distribution function that we use in the above formulas will become inaccurate. For this and some additional reasons, when the impurity

Important Formulas for Semiconductors in Equilibrium

concentration reaches about 10^{19} cm^{-3}, Equations 4.4 and 4.5 lose their validity. Under such conditions we refer to the semiconductor as *degenerate*.

An alternate set of formulas which are more useful when symmetry in the mathematical expressions is desirable can be derived from Equations 4.4 and 4.5. These are

$$n = n_i e^{(E_F - E_i)/kT} \quad (4.6)$$

$$p = n_i e^{(E_i - E_F)/kT} \quad (4.7)$$

where

$$E_i \equiv \tfrac{1}{2}(E_c + E_v) + \tfrac{1}{2}kT \ln \frac{N_v}{N_c} \quad (4.8)$$

is the intrinsic Fermi level. The intrinsic Fermi level is thus seen to be in the vicinity of the middle of the gap, displaced from it by a term which is usually very small. Consequently, for most purposes we can regard the intrinsic Fermi level to be in the middle of the energy gap.

b. The *pn* Product in Equilibrium

It is evident from Equations 4.4 and 4.5 as well as from Equations 4.6 and 4.7, that the product of electron and hole concentrations is independent of the Fermi level; hence, it is independent of the type of the semiconductor as well as of the individual electron and hole concentrations. Thus,

$$pn = n_i^2 = N_c N_v e^{-E_G/kT}. \quad (4.9)$$

This very important relationship can be derived on the basis of the law of mass action without even obtaining the individual carrier concentration equations. It always holds, provided the semiconductor is under equilibrium conditions. Thus we can employ it as the criterion for equilibrium in semiconductors.

Equation 4.9 also indicates that the intrinsic carrier concentration should depend on temperature approximately in an exponential manner $n_i \propto e^{-E_G/2kT}$, in agreement with the experimental results shown in Figure 4.4. (The actual temperature dependence is somewhat stronger because N_c and N_v themselves also increase with temperature.)

c. Space-Charge Neutrality

A semiconductor in which impurities are uniformly distributed will obey the condition of *space-charge neutrality*, which means that the net charge density ρ in any volume element of the semiconductor will be zero.

TABLE 4.1 IMPORTANT PROPERTIES OF GERMANIUM, SILICON, GALLIUM ARSENIDE, AND OF SILICON DIOXIDE AT 27°C.

	Ge	Si	GaAs	SiO_2
Atomic or molecular weight	72.60	28.09	144.63	60.08
Atoms or molecules/cm³	4.42×10^{22}	5.00×10^{22}	2.21×10^{22}	2.3×10^{22}
Crystal structure	Diamond, 8 atoms/unit cell	Diamond, 8 atoms/unit cell	Zinc-blende, 8 atoms/unit cell	Random network of SiO_4 tetrahedra. 50% covalent, 50% ionic bonding
Lattice constant (Å)	5.66	5.43	5.65	...
Density, ρ (g/cm³)	5.32	2.33	5.32	2.27
Energy gap (ev)	0.67	1.11	1.40	~8
Effective density of states conduction band N_c (cm⁻³) valence band N_v (cm⁻³)	1.04×10^{19} 6.0×10^{18}	2.8×10^{19} 1.04×10^{19}	4.7×10^{17} 7.0×10^{18}	...
Intrinsic carrier concentration n_i (cm⁻³)	2.4×10^{13}	1.45×10^{10}	$\sim 9 \times 10^{6}$	...

Important Formulas for Semiconductors in Equilibrium

Lattice (intrinsic) mobilities (cm²/v sec) electrons holes	3900 1900	1350 480	8600 250	Insulator; $\rho > 10^{16}$ Ω-cm at 300°K.
Dielectric constant	16.3	11.7	12	3.9
Breakdown field (v/μ)	~8	~30	~35	~600
Melting point (°C)	937	1415	1238	~1700
Vapor pressure (Torr)	10^{-7} at 880°C 10^{-9} at 750°C	10^{-5} at 1250°C 10^{-7} at 1050°C	1 at 1050°C 100 at 1220°C	10^{-3} at 1450°C 10^{-1} at 1700°C
Specific heat, C_p (Joule/g°C)	0.31	0.7	0.35	1.0
Thermal conductivity, k_{th} (watt/cm°C)	0.6	1.5	0.81	0.014
Thermal diffusivity $\kappa = \dfrac{k_{th}}{\rho C_p} \left(\dfrac{cm^2}{sec}\right)$	0.36	0.9	0.44	0.006
Linear coefficient of thermal expansion $\dfrac{\Delta L}{L \Delta T}\left(\dfrac{1}{°C}\right)$	5.8×10^{-6}	2.5×10^{-6}	5.9×10^{-6}	0.5×10^{-6}

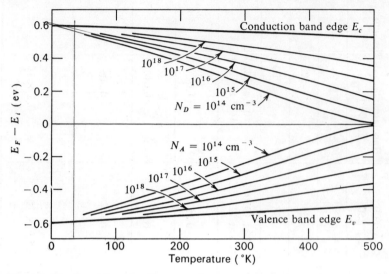

Fig. 4.7 The Fermi level in silicon as a function of temperature, for various impurity concentrations.

To obtain the net charge density, we add up all positive and negative charges. Thus, if the donors and acceptors are fully ionized,

$$\rho = q(p - n + N_D - N_A). \tag{4.10}$$

Hence, the condition of space-charge neutrality implies that

$$p - n = N_A - N_D. \tag{4.11}$$

We can combine the condition of space-charge neutrality with the equilibrium relationship, $np = n_i^2$. This leads to the concentration of electrons in an n-type semiconductor, in equilibrium,

$$n_n = \tfrac{1}{2}[N_D - N_A + \sqrt{(N_D - N_A)^2 + 4n_i^2}] \tag{4.12}$$

and the concentration of holes in a p-type semiconductor

$$p_p = \tfrac{1}{2}[N_A - N_D + \sqrt{(N_A - N_D)^2 + 4n_i^2}]. \tag{4.13}$$

We can see that when the magnitude of the net impurity concentration $|N_D - N_A|$ is much larger than the intrinsic carrier concentration n_i (this is generally the case for silicon at room temperature), the above relationships simplify to

$$n_n \doteq N_D - N_A \tag{4.14}$$

Important Formulas for Semiconductors in Equilibrium

and
$$p_p \doteq N_A - N_D. \tag{4.15}$$

Again, combining the above formulas for the concentration of majority carriers with the equilibrium relationship yields the concentration of the *minority carriers*:

$$p_n \doteq \frac{n_i^2}{N_D - N_A} \tag{4.16}$$

and

$$n_p \doteq \frac{n_i^2}{N_A - N_D}. \tag{4.17}$$

In the above formulas the subscripts refer to the *type* of the semiconductor (for example, n_n denotes the concentration of electrons in an *n*-type semiconductor). The carrier whose concentration is the larger of the two is referred to as the *majority carrier* and the other one is referred to as the *minority carrier*.

Using Equations 4.12 and 4.13 together with 4.6 and 4.7, we can calculate the position of the Fermi level within the forbidden gap as a function of temperature, for a given acceptor or donor concentration. Such calculations are shown in Figure 4.7 for silicon. Note that as the temperature increases, the Fermi level approaches the intrinsic Fermi level, i.e., the semiconductor becomes intrinsic. (This figure also indicates the slight variation of the silicon energy gap with temperature.)

Conversely, we can consider the concentration of majority carriers as a function of temperature. Figure 4.8 shows experimental measurements[2] of the electron concentration in *n*-type silicon as a function of temperature. At low temperatures the thermal energy in the crystal is not sufficient to ionize all of the donor impurities present. Thus the electron concentration is less than the donor concentration. As the temperature is increased, the condition of complete ionization, where the electron concentration equals the donor concentration, is approached. As the temperature is further increased, the electron concentration remains practically constant over a considerable temperature range. However, as the temperature is increased even further, we reach the condition where the intrinsic carrier concentration becomes comparable to the donor concentration. Beyond this temperature the semiconductor becomes intrinsic. Thus it is evident that a semiconductor may be intrinsic even if it is doped with a relatively high concentration of donors or acceptors if the temperature is high enough so that the intrinsic carrier concentration exceeds the donor or acceptor concentration. (This is evident also from the calculations shown in Figure 4.7.) The temperature at which the semiconductor becomes

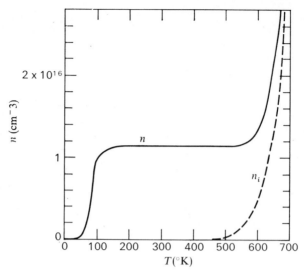

Fig. 4.8 Electron concentration in n-type silicon as a function of temperature.[2]

intrinsic of course depends on the concentration of donors or acceptors in the semiconductor.

The above formulas pertaining to semiconductors in thermal equilibrium are summarized in Table 4.2 at the end of this chapter.

4.5 TRANSPORT OF ELECTRONS AND HOLES

a. Drift

Let us consider an n-type semiconductor, with uniform donor concentration, in the absence of an applied electric field. The electrons in the semiconductor will undergo a continual random thermal motion interrupted by collisions, as illustrated in Figure 4.9a. The thermal motion leads to no net displacement of the electrons over a long enough period of time.

If an electric field is applied to the semiconductor sample, an additional velocity component will be superimposed upon the thermal motion of the carriers. This additional velocity component, called the *drift velocity*, will have a direction opposite to (for electrons) the electric field. The displacement of the electron due to this component is shown in Figure 4.9b. When we add these two components, we obtain the combined displacement of electrons, as illustrated in Figure 4.9c.

Transport of Electrons and Holes

To see what factors influence the drift velocity of electrons, we can consider the following simplified picture. The magnitude of the drift velocity at a given time t after a collision will be given by $v(t) = v(0) + at$ where $v(0)$ is the drift velocity immediately upon collision. We will take it as zero. This is equivalent to assuming that the electrons suffer collisions

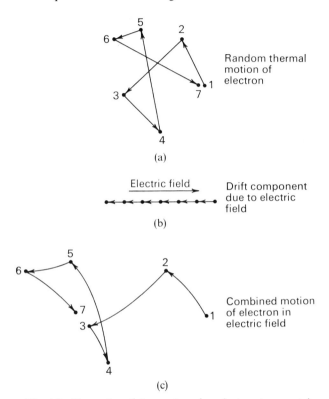

Fig. 4.9 Illustration of the motion of an electron in a crystal.

which completely randomize their motion. The magnitude of the acceleration a is given by Newton's second law as $a = q\mathscr{E}/m^*$, where m^* is the *effective mass* of the electrons in the semiconductor crystal. The effective mass is a quantity which takes the place of the mass of a free electron m in various calculations. It essentially corrects for the effect of the presence of the semiconductor crystal lattice on the behavior of the electron.

If the time interval between collisions is t_{coll}, then the average drift velocity of the electrons will be

$$\bar{v}_{\text{drift}} = \frac{q\mathscr{E}}{2m^*} t_{\text{coll}} = \mu\mathscr{E} \qquad (4.18)$$

where

$$\mu \equiv \frac{\bar{v}_{\text{drift}}}{\mathscr{E}} = \frac{qt_{\text{coll}}}{2m^*} \quad (4.19)$$

is the *mobility* of the electrons.†

This treatment assumes that the time interval between collisions t_{coll} is independent of the applied electric field. This is a reasonable assumption only so long as the drift velocity is small in comparison to the thermal velocity of carriers which is about 10^7 cm/sec for Si at room temperature.

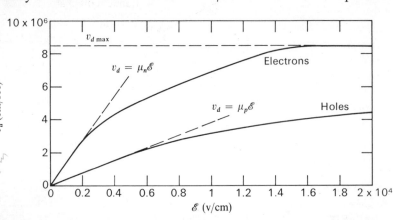

Fig. 4.10 Effect of electric field on the magnitude of the drift velocity of carriers in silicon.[3]

As the drift velocity becomes comparable to the thermal velocity, its dependence on electric field will begin to depart from the simple relationship given above. This is illustrated by the experimental measurements[3] of the drift velocity of electrons and holes in silicon as a function of the electric field, shown in Figure 4.10. Evidently an initial straight-line dependence is followed by a less rapid increase as the electric field is increased. At large enough fields, a maximum drift velocity seems to be approached.

b. Mobilities of Electrons and Holes

The time interval between collisions is determined by the various mechanisms by which the electrons or holes can lose their acquired drift velocity. The probability of a collision taking place in a unit time, $1/t_{\text{coll}}$,

† This argument is actually rather oversimplified. A more careful analysis would lead to a similar formula but without the factor 2 appearing in the denominator.

Transport of Electrons and Holes

is the sum of the probabilities of collisions due to the various such *scattering* mechanisms, i.e.,

$$\frac{1}{t_{\text{coll}}} = \frac{1}{t_{\text{coll, impurity}}} + \frac{1}{t_{\text{coll, lattice}}}$$

or

$$\frac{1}{\mu} = \frac{1}{\mu_I} + \frac{1}{\mu_L} \tag{4.20}$$

which correspond to the two most important scattering mechanisms, *impurity* and *lattice scattering*.

Impurity scattering is due to the fact that when an electron travels past a fixed charged particle, e.g., an ionized acceptor or donor, its path will be deflected by the charge on that fixed particle. The probability of impurity scattering will depend on the total concentration of ionized impurities C_T present in the crystal, i.e., the *sum* of the concentrations of negatively and positively charged ions. The mobility dominated by impurity scattering μ_I can theoretically be shown to be proportional to $T^{3/2}/C_T$.

Lattice scattering is due to the thermal vibration of the atoms of the crystal lattice which disrupts the periodicity of the lattice and thereby impedes the motion of electrons. Theoretical analysis shows that the mobility dominated by lattice scattering μ_L is proportional to $1/T^{3/2}$.

Experimentally measured[4] electron and hole mobilities in silicon at room temperature are shown in Figure 4.11 versus the total ionized impurity concentration C_T. It is seen that the mobility reaches a maximum value at low impurity concentrations corresponding to the lattice-scattering limitation, and that both electron and hole mobilities decrease with increasing impurity concentration, eventually approaching a minimum value at high concentrations. It can also be seen that the mobility of electrons is larger than the mobility of holes. This is the case in many semiconductors.

Experimental measurements[5] of the influence of temperature on the mobility of holes in silicon are shown in Figure 4.12 for two different impurity concentrations. We can distinguish two regions: at low temperatures, impurity scattering dominates and separate curves are observed for the different doping concentrations. At high temperatures, lattice scattering dominates and the impurity concentration has little effect on the mobility as evidenced by the merging of the curves. The mobility is seen to decrease with increasing temperature in this range. Experimentally, mobilities have been found to follow a $T^{-2.5}$ dependence rather than the theoretically predicted $T^{-1.5}$ dependence in the lattice scattering range.

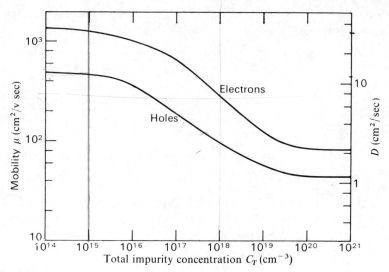

Fig. 4.11 The effect of the total ionized impurity concentration on the mobility of carriers in silicon at room temperature.[4] Also shown are the corresponding values of diffusivity.

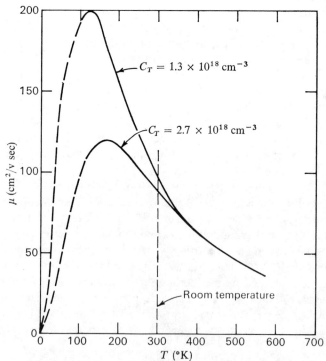

Fig. 4.12 Illustration of the effect of temperature on the mobility of carriers in silicon.[5]

c. Conduction in Homogeneous Semiconductors; Resistivity

Let us now examine the representation of the conduction process in terms of the band diagram, taking an *n*-type semiconductor as an example. Figure 4.13a shows an *n*-type semiconductor sample with no bias applied. Figure 4.13b shows the same sample with +2-volt bias applied to one of the terminals.† In the lower half of the figure we show the corresponding energy bands as a function of position along the semiconductor sample.

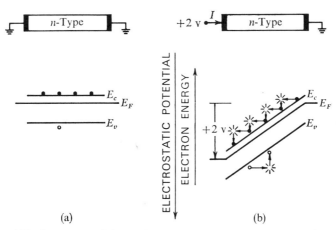

Fig. 4.13 Illustration of the conduction process in an *n*-type semiconductor.

The electrons in the conduction band will accelerate as a result of the applied field. During acceleration, they will neither gain nor lose a significant amount of their total energy so they will proceed along a more or less horizontal path in the energy band representation. While the total energy of the electrons does not change significantly during the acceleration process, they will lose potential energy, and will increase their kinetic energy at the expense of this loss in potential energy. This is evidenced by the fact that the electron trajectory takes the electron some distance above the conduction-band edge (recall that the conduction-band edge represents the potential energy of the electron).

When an electron suffers a collision, it loses some or all of its kinetic energy and imparts it to the semiconductor lattice. In this manner the kinetic energy of the electrons will be converted to heat. After the electron has lost some or all of its kinetic energy, it will again begin to accelerate

† We ignore any effects associated with the entry or exit of the electrons into and out of the semiconductor, i.e., we will ignore metal-semiconductor contact effects.

and the same process will be repeated many times. This is how the conduction process appears in the band diagram. Conduction by holes can be pictured by a similar but opposite process.

In this discussion we have considered a homogeneous semiconductor, i.e., a semiconductor in which the impurity concentration is spatially uniform. In order to keep the electron concentration spatially uniform also, we had to indicate the Fermi level in Figure 4.13 at the same distance from the conduction band at all points. Thus the Fermi level follows the conduction-band edge in a parallel fashion.

The current flowing in the semiconductor sample will be given by

$$I = q \cdot n \cdot \bar{v}_{\text{drift}} \cdot A = q n \mu_n \frac{V}{L} A \qquad (4.21)$$

where V is the voltage applied across the semiconductor sample which has a length L and a cross-sectional area A and μ_n is the electron mobility. The resistance of such a sample is given by

$$R = \rho \frac{L}{A}. \qquad (4.22)$$

Thus by comparison of the two formulas we find that the *resistivity* ρ of the *n*-type semiconductor sample is

$$\rho = \frac{1}{q \mu_n n}. \qquad (4.23)$$

Similarly, the resistivity of a *p*-type semiconductor sample is

$$\rho = \frac{1}{q \mu_p p}. \qquad (4.24)$$

In general, when both carriers are taken into consideration, the resistivity is given by

$$\rho = \frac{1}{q(\mu_n n + \mu_p p)}. \qquad (4.25)$$

The mobilities, as discussed above, depend on the total concentration of ionized impurities, hence, on the *sum* of the acceptor and donor concentrations. The electron and hole concentrations, on the other hand, depend on the *difference* of the acceptor and donor concentrations. Thus, in a general case, the resistivity must be calculated by using the mobility data given in Figure 4.11 and carrier concentrations based on Equations 4.14 or 4.15. However, in the case when only one type of impurity is present in the semiconductor, the resistivity will become a simple function of the concentration of that impurity.

The resistivity of both *p*- and *n*-type silicon at room temperature as a function of acceptor or donor concentration, respectively, is shown in

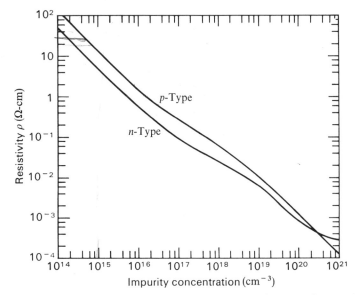

Fig. 4.14 Resistivity of silicon at room temperature as a function of acceptor or donor impurity concentration.[6]

Figure 4.14. This figure is based on an extensive survey of large numbers of measurements on samples which contain either acceptor or donor impurities.[6]

d. Diffusion

The discussion in this section has so far dealt only with the case when the electron concentration within the semiconductor sample is uniform and the electrons move under the influence of an electric field. If the electron concentration is not uniform, electrons will also *diffuse* under the influence of the concentration gradient. As in the case of ions, this will lead to an additional contribution to the expression for the flux. This contribution is proportional to the concentration gradient, and the proportionality constant is the diffusivity of electrons, D_n. The diffusivity of electrons, in turn, is related to the electron mobility by Einstein's relationship,

$$D_n = \frac{kT}{q} \mu_n. \tag{4.26}$$

Similar considerations apply to the transport of holes.

The diffusivity of electrons and holes at room temperature can be obtained from Figure 4.11 where the right-hand axis is labeled in terms of diffusivities.

READING REFERENCES

The band theory of, and electronic conduction in, solids is discussed in detail in a variety of texts on solid-state and semiconductor physics. See, for instance, Chapters 10–13 in A. J. Dekker, *Solid State Physics*, Prentice-Hall, 1957; and Chapters 1–5 in J. L. Moll, *Physics of Semiconductors*, McGraw-Hill Book Co., 1964. A tutorial treatment of band theory is given by F. Herman, "The Electronic Energy Band Structure of Silicon and Germanium," *Proc. IRE*, **43**, 1703 (1955).

For the electronic properties of semiconductors, see E. M. Conwell, "Properties of Silicon and Germanium," *Proc. IRE*, **46**, 1281 (1958); and O. Madelung, *Physics of III–V Compounds*, Wiley, 1964.

REFERENCES CITED

1. These intrinsic carrier concentrations are based on:
    a. R. N. Hall and J. H. Racette, "Diffusion and Solubility of Copper in Extrinsic and Intrinsic Germanium, Silicon, and Gallium Arsenide." *J. Appl. Phys.*, **35**, 379 (1964).
    b. F. J. Morin and J. P. Maita, "Electrical Properties of Silicon Containing Arsenic and Boron," *Phys. Rev.*, **96**, 28 (1954).
    c. F. J. Morin and J. P. Maita, "Conductivity and Hall Effect in the Intrinsic Range of Germanium," *Phys. Rev.*, **94**, 1525 (1954).
2. F. J. Morin and J. P. Maita, "Electrical Properties of Silicon Containing Arsenic and Boron," *Phys. Rev.*, **96**, 28 (1954).
3. E. J. Ryder, "Mobility of Holes and Electrons in High Electric Fields," *Phys. Rev.*, **90**, 766 (1953).
4. E. M. Conwell, "Properties of Silicon and Germanium," *Proc. IRE*, **46**, 1281 (1958).
5. G. L. Pearson and J. Bardeen, "Electrical Properties of Pure Silicon and Silicon Alloys Containing Boron and Phosphorus," *Phys. Rev.*, **75**, 865 (1949).
6. J. C. Irvin, "Resistivity of Bulk Silicon and of Diffused Layers in Silicon," *Bell System Tech. J.*, **41**, 387 (1962).

PROBLEMS

4.1 (a) Verify the formula for the intrinsic Fermi level, Equation 4.8.
    (b) Calculate the location of the intrinsic Fermi level of silicon at $-78°C$, $27°C$, and $300°C$. Is it reasonable to assume it is in the center of the forbidden gap?

4.2 Draw the energy band diagram:
    (a) at $-78°C$,
    (b) at room temperature, and
    (c) at $300°C$ for silicon doped with 10^{16} boron atoms/cm^3. Show the band edges, the intrinsic Fermi level, and the actual Fermi level. Using the intrinsic Fermi

Problems

level as a reference, label the electron energy E, and the electrostatic potential ϕ corresponding to each level.

4.3 Draw the energy band diagram corresponding to the above cases with 20 volts applied across the silicon sample. Keep the diagrams to scale. Calculate the current flow in each case if the length of the sample is 100 μ, and its cross-sectional area is 10^{-3} cm².

4.4 Prepare plots of the Fermi-Dirac distribution function at $-78°C$, room temperature, and $500°C$. Compare the three curves.

4.5 A small concentration of minority carriers is injected into a homogeneous semiconductor crystal at one point. An electric field of 10 v/cm is applied across the crystal, and this electric field moves these minority carriers a distance of 1 cm in a time 250 μsec. Determine the drift velocity and the diffusion coefficient of the minority carriers.

4.6 For an electron mobility of 500 cm²/v sec, calculate the time between collisions. For an electric field of 100 v/cm, calculate also the distance traveled by an electron between collisions. Take $m^* = m$ in these calculations.

4.7 Calculate the electron and hole concentrations, the resistivity, and the position of the Fermi level of a silicon crystal containing 1.1×10^{16} boron atoms/cm³ *and* 9×10^{15} phosphorus atoms/cm³, at $27°C$.

4.8 Calculate the Fermi level of silicon doped with 10^{16}, 10^{18}, and 10^{19} donor atoms/cm³ at room temperature, assuming complete ionization. Then, using the Fermi levels you have found, check if this assumption is justified in each case. In these calculations, take the donor level to be 0.05 ev below the conduction band edge.

4.9 Give the equilibrium electron and hole concentrations, mobilities, and resistivity for silicon at $27°C$, for each of the following impurity concentrations:
(a) 3×10^{15} boron/cm³.
(b) 1.3×10^{16} boron/cm³ $+ 1.0 \times 10^{16}$ phosphorus/cm³.
(c) 1.3×10^{16} phosphorus/cm³ $+ 1.0 \times 10^{16}$ boron/cm³.
(d) 3×10^{15} phosphorus/cm³ $+ 1.0 \times 10^{17}$ gallium/cm³ $+ 1.0 \times 10^{17}$ arsenic/cm³.

4.10 Repeat, at $300°C$. Compare the two cases and discuss.

4.11 Gold in silicon introduces an acceptor level 0.54 ev below the conduction band edge, and a donor level 0.35 ev above the valence band. What will be the state of charge (or occupation) of the gold levels in silicon doped with:
(a) High concentration of donor atoms (with respect to the gold concentration)?
(b) High concentration of acceptor atoms?
What is the effect of gold on the electron and hole concentrations, respectively?

4.12 Using the information given in the previous problem, determine the charge state of the gold levels and the position of the Fermi levels in a silicon crystal containing gold atoms only. Is the crystal p- or n-type?

TABLE 4.2
IMPORTANT FORMULAS IN SEMICONDUCTOR PHYSICS
Complete ionization of impurities
Thermal equilibrium

Charge neutrality	$\rho = q(p - n + N_D - N_A) = 0$
Equilibrium condition	$pn = n_i^2$
Fermi-Dirac distribution function	$f(E) = \dfrac{1}{1 + e^{(E-E_F)/kT}}$
Carrier concentrations in non-degenerate semiconductors:	$n = N_c e^{-(E_c-E_F)/kT} = n_i e^{(E_F-E_i)/kT}$ $p = N_v e^{-(E_F-E_v)/kT} = n_i e^{(E_i-E_F)/kT}$
In the extrinsic case, $\|N_D - N_A\| \gg n_i$:	$n_n \doteq N_D - N_A \qquad p_p \doteq N_A - N_D$ $p_n \doteq \dfrac{n_i^2}{N_D - N_A} \qquad n_p \doteq \dfrac{n_i^2}{N_A - N_D}$

- INJECTION
- KINETICS OF THE RECOMBINATION PROCESS
- LIFETIME FOR LOW-LEVEL INJECTION
- SURFACE RECOMBINATION
- ORIGIN OF RECOMBINATION-GENERATION CENTERS

5

Semiconductors under Non-Equilibrium Conditions

Most semiconductor devices operate under non-equilibrium conditions, i.e., under conditions in which the carrier concentration product pn differs from its equilibrium value, n_i^2. The performance of many semiconductor devices is determined by their tendency to return to equilibrium. In this chapter we derive and examine the quantities which characterize this tendency: *lifetime, diffusion length,* and *surface recombination velocity*.

We begin by discussing the concept of injection of excess carriers. The process of return to equilibrium then is considered through examples of two types: transient cases, and steady-state cases in which the distribution of excess carriers is non-uniform. We show that both types of cases can be characterized by the lifetime of the excess carriers, their diffusion length, and the surface recombination velocity. In order to relate these quantities to characteristics of the semiconductor, we then study the mechanism of recombination, both in the bulk and at the surface. Finally, we discuss the physical origin of bulk and surface recombination centers.

5.1 INJECTION

Let us consider non-equilibrium situations in which the condition $np = n_i^2$ is violated. Accordingly, we can distinguish between two types of deviation from equilibrium. In the first type, when $np > n_i^2$, we talk of *injection* of excess carriers. In the second, when $np < n_i^2$, we talk of the *extraction* of carriers from the semiconductor.

a. Injection Level

The meaning of *injection level* is illustrated in Figure 5.1. Here we indicate the donor concentration $N_D = 10^{16}$ cm^{-3} and the majority and minority carrier concentrations in an *n*-type semiconductor under equilibrium, and low- and high-level injection conditions.† As we have seen in

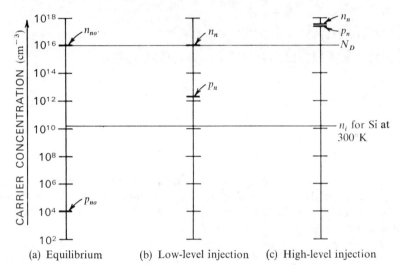

(a) Equilibrium (b) Low-level injection (c) High-level injection

Fig. 5.1 Illustration of the concentration of electrons and holes in an *n*-type semiconductor in equilibrium and under low- and high-level injection conditions.

Chapter 4, under equilibrium conditions the product of majority and minority carrier concentrations must equal n_i^2, or approximately 10^{20} cm^{-6} in silicon at room temperature. The majority carrier concentration approximately equals the donor concentration. Hence,

$$n_{no} = 10^{16} \text{ cm}^{-3} \quad \text{and} \quad p_{no} = 10^4 \text{ cm}^{-3}$$

as shown in Figure 5.1a. In this notation the first subscript refers to the *type* of the semiconductor. The subscript *o* indicates an equilibrium quantity. Thus n_{no} and p_{no} denote the electron and hole concentrations, respectively, in an *n*-type semiconductor in equilibrium.

Let us now consider the case when we somehow introduce excess carriers of both types into the semiconductor, in equal concentrations in order to

† Throughout this chapter we will use an *n*-type semiconductor as an example. All results, however, are equally applicable to a *p*-type semiconductor.

Injection

preserve space-charge neutrality. In the example shown in Figure 5.1b we have increased the minority carrier concentration a hundred-million fold, from 10^4 to 10^{12} cm^{-3}. Although at the same time we have also added approximately 10^{12} cm^{-3} majority carriers to the semiconductor, this concentration of excess electrons is negligibly small in comparison to the concentration of electrons already present in the *n*-type semiconductor. Thus, while the magnitude of the change in the concentration of electrons is the same as the change in the concentration of the holes, the percentage change in electron concentration is negligible. This condition, in which the excess carrier concentration is negligibly small in comparison to the doping concentration, i.e., $\Delta n = \Delta p \ll N_D$, is referred to as *low-level injection*.

For comparison, we also illustrate the case of *high-level injection* in Figure 5.1c. In high-level injection the injected excess carriers are in larger, or at least in comparable, concentrations to the concentration of the donor ions. Thus in this case the percentage change in majority carrier concentration is no longer negligible. Even though high-level injection is often encountered in semiconductor device operation, because of the complexities involved in its treatment we shall consider only low-level injection.

b. Return to Equilibrium

Whenever the carrier concentrations are disturbed from their equilibrium values they will attempt to return to equilibrium. In the case of injection of excess carriers, return to equilibrium is through *recombination* of the injected minority carriers with the majority carriers. In the case of extraction of carriers, return to equilibrium is through the process of *generation* of electron-hole pairs.

We now develop the parameter which characterizes the rate of return to equilibrium. This parameter, as well as certain other concepts of the utmost importance in semiconductor device operation, can best be developed by considering specific examples, starting with the simplest case.

Decay of Injected Carriers. Consider a uniformly illuminated semiconductor slice as shown in the inset to Figure 5.2. We assume that the light is so energetic that it creates electron-hole pairs within the semiconductor. Furthermore, we assume that the light is absorbed uniformly throughout the semiconductor sample, resulting in a uniform rate of generation G_L of electron-hole pairs per unit volume throughout the crystal. The resulting spatially uniform distribution of minority carriers

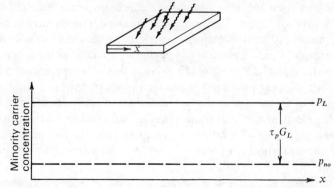

Fig. 5.2 Illustration of the steady-state minority carrier distribution in a uniformly illuminated semiconductor sample, in the absence of surface recombination.

in the semiconductor specimen is illustrated in Figure 5.2. Because of the increased generation rate resulting from illumination, the concentration of minority carriers will build up to a steady-state level high enough to make the rate of recombination of the carriers just equal to the rate of generation.

To calculate the steady-state minority carrier concentration reached under illumination we first note that the time rate of increase of minority carrier concentration dp_n/dt equals the *total* generation rate of minority carriers minus the *total* recombination rate, or

$$\frac{dp_n}{dt} = G_L + G_{th} - R \tag{5.1}$$

where G_L is the rate of generation due to the absorbed light, G_{th} is the rate of generation in dark, due to thermal mechanisms, and R is the total rate of recombination (all per unit time and unit volume). It is convenient to define the *net rate* of *recombination*, $U \equiv R - G_{th}$, and rewrite the above equation to give

$$\frac{dp_n}{dt} = G_L - U. \tag{5.2}$$

By this definition the net rate of recombination $U = 0$ in the steady state if there is no injection by light or other means.

Before we can solve this equation, we need a relationship between U and the minority carrier concentration. We *assume* the simplest possible relationship, namely that U is proportional to the excess minority carrier concentration, or

$$U = \frac{1}{\tau_p}(p_n - p_{no}). \tag{5.3}$$

Injection

This assumption has the correct feature that $U = 0$ in equilibrium. The constant of proportionality $1/\tau_p$ will have to be determined from a study of the mechanism of the recombination process. The constant τ_p (or τ_n for a *p*-type semiconductor) is referred to as the *lifetime* of the excess minority carriers. The form of this equation is similar to others we have used earlier in the treatment of solid-state processes. In all of them we have assumed that a rate is proportional to a driving force—a measure of the deviation from equilibrium.

Combining the above relationships, we get the differential equation describing the minority carrier concentration in the semiconductor as a function of time,

$$\frac{dp_n}{dt} = G_L - \frac{p_n - p_{no}}{\tau_p}. \tag{5.4}$$

In steady state, $dp_n/dt = 0$. Thus the steady-state concentration of minority carriers during illumination is given by

$$p_L = p_{no} + \tau_p G_L. \tag{5.5}$$

If the light is now turned off, i.e., $G_L = 0$, the excess minority carrier concentration will begin to decay. This decay is described by the solution of the differential equation,

$$\frac{dp_n}{dt} = -\frac{p_n - p_{no}}{\tau_p} \tag{5.6}$$

subject to the initial condition,

$$p_n(0) = p_L. \tag{5.7}$$

The solution is given by

$$p_n(t) = p_{no} + (p_L - p_{no})e^{-t/\tau_p}. \tag{5.8}$$

This solution is illustrated for various values of the lifetime τ_p in Figure 5.3.

Surface Recombination in Steady State. Let us again consider a semiconductor slice which is uniformly illuminated as in the previous case. Let us further assume that the recombination rate of the excess carriers is enhanced at one side of this slice, as illustrated in the inset to Figure 5.4. Because of the enhanced rate of recombination at the surface corresponding to the plane $x = 0$, the concentration of the excess minority carriers will be lower at this plane than in the body of the semiconductor sample. As a result, minority *and* majority carriers will flow to this surface and recombine there. Because one electron recombines with one hole, the flux of holes to the surface F_p will precisely equal the flux of electrons F_n. Thus there will be *no net current* flowing.

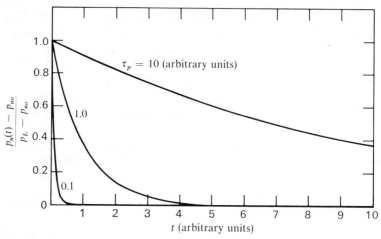

Fig. 5.3 The decay of excess minority carrier concentration as a function of time in the transient case, for various values of the lifetime.

In this problem both minority and majority carrier distributions vary spatially. Thus the hole distribution is described by a solution of the transport equation,

$$\frac{\partial p_n}{\partial t} = -\frac{\partial F_p}{\partial x} + G_L - U \tag{5.9}$$

where F_p denotes the flux of holes.

Equation 5.9 can be derived by considerations exactly like those we employed in the derivation of the transport equation in Chapter 3, but

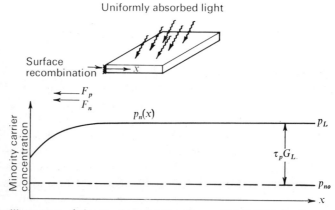

Fig. 5.4 Illustration of the steady-state minority carrier distribution in a uniformly illuminated semiconductor sample, with surface recombination.

Injection

including the generation and recombination terms into the balance over the control-element in Figure 3.6. The flux of holes is given by

$$F_p = -D_p \frac{\partial p_n}{\partial x} + \mu_p \mathscr{E} p_n \qquad (5.10)$$

where D_p and μ_p are the diffusivity and mobility of holes, respectively (see Chapters 3 and 4).

This problem would be greatly simplified if we could neglect the drift term, i.e., the second term of the flux equation. To show that for *minority carriers in low-level injection the drift term is in fact negligible in comparison to the diffusion term*, we recall that the flux of holes to the surface equals the flux of electrons, which is given by

$$F_n = -D_n \frac{\partial n_n}{\partial x} - \mu_n \mathscr{E} n_n \qquad (5.11)$$

where D_n and μ_n are the diffusivity and mobility of electrons, respectively. From the equality of the two fluxes $F_p = F_n$, and from the requirement of space-charge neutrality due to which $\partial n_n/\partial x = \partial p_n/\partial x$, we obtain the expression for the electric field,

$$\mathscr{E} = -\frac{(D_n - D_p)(\partial p_n/\partial x)}{\mu_p p_n + \mu_n n_n}. \qquad (5.12)$$

In low-level injection $p_n \ll n_n$; thus the drift term in Equation 5.10 is given by

$$\mu_p \mathscr{E} p_n \doteq -(D_n - D_p) \frac{\partial p_n}{\partial x} \frac{\mu_p}{\mu_n} \frac{p_n}{n_n} = -D_p \frac{\partial p_n}{\partial x} \left(1 - \frac{D_p}{D_n}\right) \frac{p_n}{n_n}. \qquad (5.13)$$

Since $p_n \ll n_n$, this term is clearly negligible in comparison to the diffusion term.

In contrast, it can be readily shown by using a similar argument that the drift term for majority carriers is not negligible in comparison to the diffusion term. Since diffusion and drift contributions to the majority carrier flux are of opposite sign, the net flux of majority carriers to the surface is given by their difference.

A graphical illustration of the relative magnitudes of the four flux terms in low- and in high-level injection is given in Figure 5.5. It is evident that whereas in high-level injection all four flux terms are of comparable magnitude, in low-level injection only the diffusion term provides an important contribution to the flux of minority carriers.

Although the above argument is for a case with no current flowing, this result is of general validity and greatly simplifies the study of the motion of minority carriers in low-level injection. Thus the transport of minority

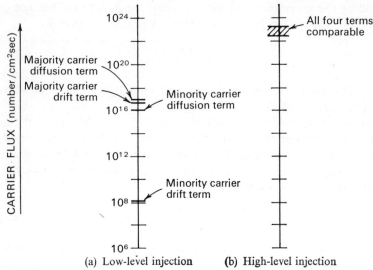

Fig. 5.5 Illustration of the relative magnitudes of the diffusion and drift terms for the case of steady-state surface recombination.

carriers is described simply by the diffusion equation with the generation and recombination terms added,

$$\frac{\partial p_n}{\partial t} = D_p \frac{\partial^2 p_n}{\partial x^2} + G_L - \frac{p_n - p_{no}}{\tau_p}. \quad (5.14)$$

Returning now to the problem of surface recombination, in steady state $\partial p_n/\partial t = 0$. We then seek a solution to the resulting ordinary differential equation subject to the boundary conditions

$$p_n(\infty) = p_L = p_{no} + \tau_p G_L \quad (5.15)$$

and

$$D_p \frac{\partial p_n}{\partial x}\bigg|_{x=0} = s_p[p_n(0) - p_{no}]. \quad (5.16)$$

The second boundary condition states that the minority carriers which reach the surface recombine there. As in the case of bulk recombination, we assume that the recombination rate at the surface is proportional to the concentration of excess minority carriers $(p_n - p_{no})$ there. The proportionality constant s_p, which has the units centimeter per second is called the *surface recombination velocity*. The solution of this boundary value problem is

$$p_n(x) = p_L - (p_L - p_{no}) \frac{s_p \tau_p / L_p}{1 + s_p \tau_p / L_p} e^{-x/L_p} \quad (5.17)$$

where $L_p \equiv \sqrt{D_p \tau_p}$ is called the *diffusion length* of minority carriers.

Injection

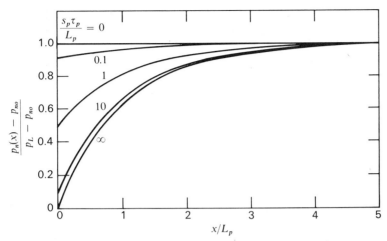

Fig. 5.6 The steady-state distribution of excess minority carriers for various relative values of surface recombination velocity.

This solution is shown in Figure 5.6 for various relative values of surface recombination velocity s_p. In the limit when the surface recombination velocity $s_p \to 0$, the solution reduces to the case illustrated in Figure 5.2. In the other limit, when $s_p \to \infty$, the minority carrier concentration at the surface approaches its equilibrium value. The distribution of minority carriers for this case reduces to

$$p_n(x) = p_L - (p_L - p_{no})e^{-x/L_p}. \qquad (5.18)$$

This distribution is shown in Figure 5.7 for various values of L_p.†

Steady-State Injection from One Boundary. A case corresponding to a different physical situation is illustrated in Figure 5.8. Here we illuminate the semiconductor sample from one side only, and in such a way that all of the light is absorbed in a very thin layer next to the semiconductor surface. Thus in this thin layer a large excess minority carrier concentration is set up. The excess minority carriers diffuse inward into the undisturbed body of the semiconductor from this surface layer as illustrated in Figure 5.8. Since no net current is entering the semiconductor sample, the majority carrier concentration distribution must again be such that the net flux of electrons and of holes will be in the same direction and equal in magnitude.

† It is interesting to note how similar this problem is to diffusion of impurities subject to external rate-limitation which we considered in Chapter 3. Accordingly, these two limiting cases could be called 'surface-recombination controlled' and 'diffusion controlled,' respectively.

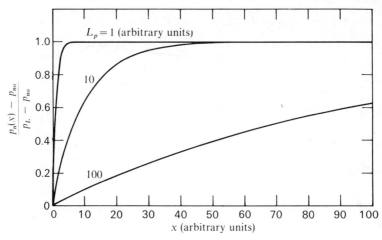

Fig. 5.7 The steady-state distribution of excess minority carriers for various values of the diffusion length, for $s_p \to \infty$.

The steady-state minority carrier distribution within the bulk of the semiconductor sample must satisfy the equation

$$D_p \frac{d^2 p_n}{dx^2} - \frac{p_n - p_{no}}{\tau_p} = 0 \tag{5.19}$$

subject to the boundary conditions,

$$p_n(0) = \text{constant, function of light intensity} \tag{5.20}$$

and

$$p_n(\infty) = p_{no}. \tag{5.21}$$

The solution is

$$p_n(x) = p_{no} + [p_n(0) - p_{no}]e^{-x/L_p} \tag{5.22}$$

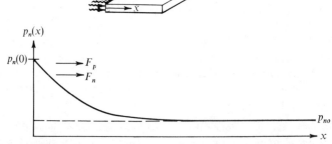

Fig. 5.8 Illustration of the minority carrier distribution in a sample illuminated on one side with non-penetrating light.

Kinetics of the Recombination Process

which is shown in Figure 5.9 as a function of distance, for three different values of the diffusion length L_p.

Note the complete analogy between the steady-state distribution of excess minority carriers as a function of distance, given by Equation 5.22 and shown in Figure 5.9, and the time-wise decay of the excess minority carrier concentration, given by Equation 5.8 and shown in Figure 5.3. In each case there is a disturbance of minority carrier concentration. In one case, the disturbance occurs at a given point in space; in the other, at a given point in time. With increasing distance, or with a passage of time,

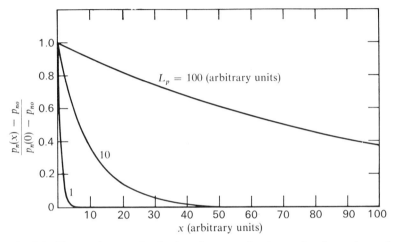

Fig. 5.9 The steady-state distribution of excess minority carriers for various values of the diffusion length.

the equilibrium minority carrier concentration is then approached. In each case the rate of approach to equilibrium is related to the same parameter, the lifetime (or, equivalently, the diffusion length) of minority carriers.

5.2 KINETICS OF THE RECOMBINATION PROCESS

In the previous section we have seen that the rate of return of the minority carrier distribution to equilibrium—whether such return takes place with passing time or with increasing distance—depends on the rate constant characterizing the recombination process $1/\tau_p$. This resulted from our assumption that the net rate of recombination in the bulk of an *n*-type semiconductor can be described by the linearized expression

$$U = (1/\tau_p)(p_n - p_{no}).$$

In this section we attempt to relate the lifetime τ_p to the physical characteristics of the semiconductor by considering the mechanisms of the recombination process.

a. Band-to-Band Recombination

When electrons in the conduction band and holes in the valence band recombine directly, we talk of a *band-to-band* recombination process which is illustrated schematically in Figure 5.10. Here we indicate generation

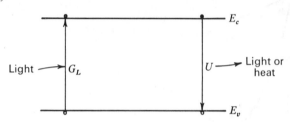

Fig. 5.10 The band-to-band recombination process.

due to light absorbed in the semiconductor, which has a rate G_L, and the net band-to-band recombination, which has a rate U. In such a recombination process the electron must lose energy of the order of the energy gap. This energy can be emitted in the form of light or in the form of heat.

We would expect the rate of the band-to-band recombination process to be proportional to both the concentrations of electrons and of holes. Thus for an *n*-type semiconductor,

$$R = \alpha n_n p_n, \tag{5.23}$$

where α is a proportionality constant.

In equilibrium, $R = G_{th} = \alpha n_{no} p_{no}$. In low-level injection, the majority carrier concentration does not change significantly, i.e., $n_n \doteq n_{no}$. Thus

$$U = R - G_{th} \doteq \alpha n_{no}(p_n - p_{no}). \tag{5.24}$$

Hence, the lifetime for the band-to-band recombination process becomes

$$\tau_p = \frac{1}{\alpha n_{no}}. \tag{5.25}$$

Although in some semiconductors, e.g., gallium arsenide, band-to-band recombination can be very important, in germanium and in silicon the detailed structure of the energy bands makes this process extremely unlikely. In fact, early work with both germanium and silicon indicated that the lifetime is extremely structure sensitive in such materials, i.e., it is sensitive to the method employed in the preparation of the semiconductor samples. We would expect the rate of the band-to-band recombination process to be dependent only on the band structure of the semiconductor. Thus the fact that the method of sample preparation has an effect on the lifetime indicates that the recombination process involves *imperfections* or *impurities* in the semiconductor whose concentration could indeed depend on the fabrication technique.

We now consider the recombination-generation process taking place through the action of such imperfections.

b. Recombination-Generation through Intermediate Centers

Imperfections within the semiconductor can disrupt the perfect periodicity of the crystal lattice, and as a result can introduce energy levels into the forbidden gap much as donor and acceptor impurities do. These energy levels then act as "stepping-stones" in the transition of electrons and holes between the conduction and valence bands. Because the probability of transitions depends on the size of the step, imperfections can make such transitions more probable and, therefore, can exert a drastic influence on the lifetime in the semiconductor.

The theory of the recombination-generation process taking place through the action of such intermediate energy-level *recombination-generation centers* has been worked out by Hall, and by Shockley and Read.[1] This theory has been remarkably successful in explaining a wide variety of phenomena in many semiconductors and semiconductor devices. Because of this we now consider it in some detail.

The various steps that occur in the recombination and generation process through intermediate-level centers are shown in Figure 5.11. In particular, we illustrate the state of the center before and after each of the four basic processes takes place. The arrows in this figure designate the transition of the electron during the particular process. This illustration is for the case of a center with a single energy level which can have two charge states: negative and neutral.

Process (a) is the *capture of an electron* from the conduction band by the center. Process (b) is the reverse process—*the emission of an electron* from the center into the conduction band. Process (c) is the *capture of a hole*

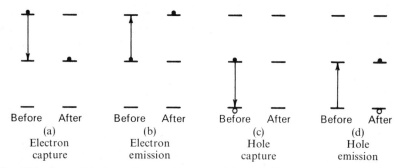

Fig. 5.11 Recombination and generation through intermediate centers. Arrows designate direction of electron transition.

from the valence band by a center. This process can also be described as the transition of an electron from the center into the valence band. Finally, process (d) is the *emission of a hole* from a center into the valence band. This can also be described as the transition of an electron from the valence band to the center, leaving behind a hole in the valence band.

Let us now consider the rates of these individual processes. The rate of electron capture—process (a)—should be proportional to the concentration of free electrons in the conduction band and also to the concentration of centers which are *not occupied* by electrons. This is because only one electron can occupy a given center; thus once a center is occupied by an electron it cannot capture another one. If the concentration of centers in the semiconductor is N_t, the concentration of unoccupied centers is given by $N_t(1-f)$ where f designates the probability of occupation of a center by an electron.† In equilibrium,

$$f = \frac{1}{1 + e^{(E_t - E_F)/kT}} \tag{5.26}$$

where E_t is the energy level of the center and E_F is the Fermi level.

Thus the rate of process (a) will be given by

$$r_a \propto nN_t(1-f). \tag{5.27}$$

We designate the proportionality constant by the product $v_{th}\sigma_n$, i.e.,

$$r_a = v_{th}\sigma_n nN_t(1-f). \tag{5.28}$$

Here v_{th} is the thermal velocity of the carriers, $v_{th} \equiv \sqrt{3kT/m} \simeq 10^7$ cm/sec at room temperature. The quantity σ_n can be interpreted as follows. In order to be captured, the electron must come to the physical vicinity of

---

† The subscript t has been traditionally employed to denote quantities pertaining to recombination-generation centers. It originates from the term *trap* which we shall not use in order to avoid confusion.

Kinetics of the Recombination Process

the center. The *capture cross-section* of the center σ_n is a measure of how close the electron has to come to the center to be captured.† We might expect that the capture cross section would be of the order of atomic dimensions, i.e., of the order of 10^{-15} cm².

The rate of electron emission—process (b)—will be proportional to the concentration of centers which are occupied by electrons, i.e., $N_t f$. Thus,

$$r_b = e_n N_t f. \quad (5.29)$$

The proportionality constant e_n, called the *emission probability*, is the probability of a jump from an occupied center into the conduction band. Thus it depends on the density of unoccupied states within the conduction band and also on the location of the center within the forbidden gap. Intuitively, we would expect that if the center is shallow, i.e., if it is close to the conduction-band edge, this jump-probability and hence e_n would be large, and vice versa. We will see later that the theory indeed bears out this guess.

The rate of capture of holes—process (c)—will by analogy to process (a) be given by

$$r_c = v_{th} \sigma_p p N_t f. \quad (5.30)$$

Since capture of holes by a center corresponds to the transition of an electron from a center to the valence band, this process is proportional to the concentration of centers occupied by electrons, $N_t f$.

Conversely, the rate of hole emission—process (d)—will be given by

$$r_d = e_p N_t (1 - f) \quad (5.31)$$

where e_p is the emission probability of holes and depends on factors analogous to those that enter into determining e_n.

First we would like to evaluate the emission probabilities, e_n and e_p. To do this, we consider the *equilibrium* case: without any external generation mechanism acting, i.e., when $G_L = 0$. In this case, the rates of the two processes through which transition into and out of the conduction band takes place must be equal. Thus $r_a = r_b$. Substituting the proper equations for these rates and recalling from Chapter 4 that the electron concentration in equilibrium is given by

$$n = N_c e^{-(E_c - E_F)/kT} = n_i e^{(E_F - E_i)/kT},$$

we obtain

$$e_n = v_{th} \sigma_n N_c e^{-(E_c - E_t)/kT} = v_{th} \sigma_n n_i e^{(E_t - E_i)/kT}. \quad (5.32)$$

† Strictly speaking, we should base the value of the thermal velocity on the *effective mass* of the carrier whose capture we are considering rather than on the mass of a free electron, m. However, for simplicity, we include the correction factor, $\sqrt{m/m_{\text{eff}}}$, with the respective capture cross sections.

Note that, in agreement with our intuitive argument, the emission probability of electrons e_n indeed increases exponentially as the center energy level E_t approaches the conduction band edge E_c.

Similarly, in equilibrium the two processes by which holes enter and leave the valence band, processes (c) and (d), must have identical rates. Thus $r_c = r_d$. Again substituting the appropriate equations and noting from Chapter 4 that the concentration of holes in equilibrium is given by

$$p = N_v e^{-(E_F - E_v)/kT} = n_i e^{(E_i - E_F)/kT},$$

we obtain

$$e_p = v_{th} \sigma_p N_v e^{-(E_t - E_v)/kT} = v_{th} \sigma_p n_i e^{(E_i - E_t)/kT}. \tag{5.33}$$

Again we note that the emission probability e_p increases exponentially as the center level E_t approaches the edge of the valence band E_v.

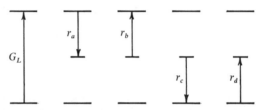

Fig. 5.12 Transitions taking place under non-equilibrium conditions.

Let us now consider the situation pertaining to non-equilibrium conditions, such as, for instance, the case of a uniformly illuminated semiconductor with a uniform generation rate per unit volume G_L. The transitions taking place under such conditions are illustrated in Figure 5.12. Note that in addition to the processes (a), (b), (c), and (d) of Figure 5.11, electrons now also leave the valence band and enter into the conduction band through the action of the illumination process. In *steady state*, the rate by which electrons enter the conduction band will equal the rate by which electrons leave the conduction band. Thus

$$\frac{dn_n}{dt} = G_L - (r_a - r_b) = 0. \tag{5.34}$$

Similarly, in steady state the rate by which holes leave the valence band equals the rate by which holes enter the valence band and, therefore,

$$\frac{dp_n}{dt} = G_L - (r_c - r_d) = 0. \tag{5.35}$$

Equations 5.34 and 5.35, of course, reduce to the equilibrium case when we set $G_L = 0$. (*Note that steady state does not imply equilibrium!*) For

Kinetics of the Recombination Process

steady-state non-equilibrium conditions we can eliminate G_L between the above expressions and write

$$r_a - r_b = r_c - r_d. \tag{5.36}$$

By substituting the proper rates into Equation 5.36 we can solve for the occupancy factor f of the centers under a given non-equilibrium condition in terms of the electron and hole concentrations. Note that neither the expression for f, Equation 5.26, nor those for n and p are meaningful under non-equilibrium conditions since the concept of the Fermi level is only valid in equilibrium. The electron and hole concentrations instead will be dependent on the injection level, i.e., on G_L, thereby making f also depend on the injection level. Thus,†

$$f = \frac{\sigma_n n + \sigma_p N_v e^{-(E_t - E_v)/kT}}{\sigma_n[n + N_c e^{-(E_c - E_t)/kT}] + \sigma_p[p + N_v e^{-(E_t - E_v)/kT}]} \tag{5.37}$$

or, in an alternate form,

$$f = \frac{\sigma_n n + \sigma_p n_i e^{(E_i - E_t)/kT}}{\sigma_n[n + n_i e^{(E_t - E_i)/kT}] + \sigma_p[p + n_i e^{(E_i - E_t)/kT}]}. \tag{5.38}$$

We can substitute these values of f into the rates of the individual processes and obtain the steady-state net rate of recombination through the action of intermediate centers U,

$$U = r_a - r_b = r_c - r_d$$

$$= \frac{\sigma_p \sigma_n v_{th} N_t[pn - n_i^2]}{\sigma_n[n + N_c e^{-(E_c - E_t)/kT}] + \sigma_p[p + N_v e^{-(E_t - E_v)/kT}]} \tag{5.39}$$

or, alternatively,

$$U = \frac{\sigma_p \sigma_n v_{th} N_t[pn - n_i^2]}{\sigma_n[n + n_i e^{(E_t - E_i)/kT}] + \sigma_p[p + n_i e^{(E_i - E_t)/kT}]}. \tag{5.40}$$

To see the principal features of this extremely important relationship, let us consider a special case when the capture cross sections for electrons and for holes are equal, i.e., $\sigma_p = \sigma_n = \sigma$. In this case Equation 5.40 reduces to

$$U = \sigma v_{th} N_t \frac{pn - n_i^2}{n + p + 2n_i \cosh\left(\dfrac{E_t - E_i}{kT}\right)}. \tag{5.41}$$

We can identify the "driving force" for recombination as $pn - n_i^2$, which is in fact the deviation from the equilibrium condition. The "resistance"

† We assume here that the illumination does not change the rates of the emission and capture processes other than through the changed carrier concentrations.

for this recombination process increases with n and with p, thus that part of the resistance will be smallest when the sum $(n + p)$ is at its minimum value.

The third term in the denominator increases as E_t moves away from the middle of the energy gap E_i and approaches either the conduction-band or the valence-band edge. In such a case one of the emission processes becomes increasingly probable and this reduces the effectiveness of the recombination center. This is because after an electron is captured by the center, a hole must be captured by it next in order to complete the recombination process. If, however, the energy level of the center is very near the conduction-band edge, it will be more likely to re-emit the captured electron into the conduction band, thereby preventing the completion of the recombination process. (A similar argument holds for centers near the valence-band edge.)

Thus a recombination center is most effective if the two emission probabilities are about the same, i.e., when its energy level is near the middle of the energy gap. In the next chapter we will see that such a center is also most effective in *generating* carriers. We can summarize this argument in simple terms by recalling what we have said earlier: that the centers provide "stepping stones" between the conduction and valence bands. For a stepping stone to be most effective it should halve the total distance between the two bands.

Let us now consider special forms of Equations 5.39 and 5.40 applicable to the specific cases we have studied in the earlier part of this chapter.

5.3 LIFETIME IN LOW-LEVEL INJECTION

In the various cases involving recombination of injected excess carriers in an n-type semiconductor, we have assumed that the net recombination rate per unit volume is given by the relationship $U = (p_n - p_{no})/\tau_p$. We can now apply Equation 5.40 to an n-type semiconductor in low-level injection. Under these conditions, $n_n \gg p_n$. Furthermore, $n_n \gg n_i e^{(E_t - E_i)/kT}$ for centers that are efficient recombination centers, i.e., that are not too near the conduction-band edge. Thus U can be approximated by

$$U = \frac{\sigma_p \sigma_n v_{th} N_t [n_n p_n - n_i^2]}{\sigma_n n_n} = \sigma_p v_{th} N_t [p_n - p_{no}]. \quad (5.42)$$

Accordingly the lifetime of holes in low-level injection in an n-type semiconductor is

$$\tau_p = \frac{1}{\sigma_p v_{th} N_t}. \quad (5.43)$$

Lifetime in Low-Level Injection

This and other important formulas relating to semiconductors under non-equilibrium conditions are summarized in Table 5.1 at the end of this chapter.

Note that the lifetime is independent of the concentration of electrons. This is because there is a great abundance of electrons in an n-type semiconductor. Thus, as soon as a hole is captured by a center, an electron will immediately be captured by the same center and the recombination process thereby completed. In other words, the *rate-limiting step in the recombination process is the capture of the minority carrier.*

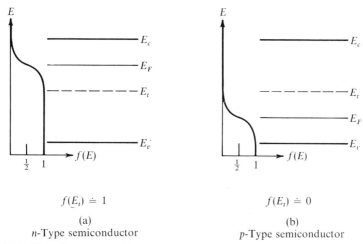

$$f(E_t) \doteq 1 \qquad\qquad f(E_t) \doteq 0$$

(a)   (b)
n-Type semiconductor p-Type semiconductor

Fig. 5.13 The charge-state of midgap centers in n- and p-type semiconductors, in equilibrium.

This can also be seen in another way. If the recombination rate will be limited by the rate of capture of the injected holes, we would expect U to equal the rate of process (c). However, for the rate of process (c) to lead to Equation 5.42, f must approximately equal 1. This means that in an n-type semiconductor in low-level injection most of the centers will be occupied by electrons. (Of course, this is what we would expect in an n-type semiconductor in equilibrium for centers near the middle of the forbidden gap, since then $E_F > E_t$ as illustrated in Figure 5.13a.) Thus, on the average, the centers are occupied by electrons and are waiting to capture holes. When one does capture a hole, it immediately captures an electron again and is occupied by an electron for a longer period of time until it again captures a hole, and so on.

A similar argument for a p-type semiconductor leads to the rate of recombination of excess minority carriers in low-level injection as

$$U = \sigma_n v_{th} N_t [n_p - n_{po}]. \tag{5.44}$$

Thus the lifetime of electrons in a p-type semiconductor is given by

$$\tau_n = \frac{1}{\sigma_n v_{th} N_t}. \qquad (5.45)$$

Comparison of Equation 5.44 with the rate of electron capture, process (a), shows that the probability of occupation of a near-midgap center by an electron in a p-type semiconductor is approximately zero, just as we would expect for a p-type semiconductor in equilibrium (see Figure 5.13b).

5.4 SURFACE RECOMBINATION

So far we have considered only recombination at centers distributed uniformly within the semiconductor. We now consider what happens if we have an increased concentration of recombination centers in a thin layer of thickness x_1 near the surface of the semiconductor sample, as illustrated in Figure 5.14. In such a case we expect that the recombination rate U will be enhanced near the surface with the result that the excess carrier density will be smaller there. In an attempt to even out the differences in excess carrier concentration in this surface layer and in the rest of the semiconductor, carriers will diffuse from the body of the semiconductor.

The fluxes of carriers that flow to this region of enhanced recombination rate supply all the carriers that recombine in this region. The total number of carriers recombining in the surface layer per unit area and unit time is given by

$$U_s = \sigma_p v_{th} N_t^* x_1 [p_n(0) - p_{no}] \qquad (5.46)$$

where $p_n(0)$ designates the average minority carrier concentration in the surface layer, and N_t^* is the concentration (per cubic centimeter) of centers there. Since the flux of minority carriers to this region must equal U_s, we arrive at the condition

$$D_p \left.\frac{\partial p_n}{\partial x}\right|_{x=0} = \sigma_p v_{th} N_t^* x_1 [p_n(0) - p_{no}]. \qquad (5.47)$$

Note that the form of this condition is identical to Equation 5.16:

$$D_p \left.\frac{\partial p_n}{\partial x}\right|_{x=0} = s_p [p_n(0) - p_{no}].$$

By comparing Equations 5.47 and 5.16 we see that the surface recombination velocity of holes s_p is given by

$$s_p = \sigma_p v_{th} N_{st} \qquad (5.48)$$

Surface Recombination

where we replaced ($N_t^* x_1$) with N_{st}, the total number of centers (per unit surface area) within the boundary region.

Thus it is evident that surface recombination can be considered as a special case of bulk recombination for a high density of centers distributed within a very thin region near a surface.

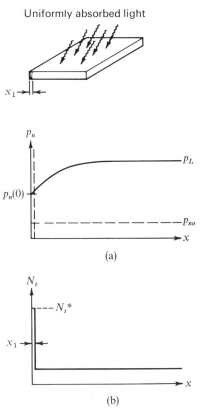

Fig. 5.14 Enhanced recombination rate near a surface due to a high center density, and the resulting distribution of excess minority carriers. (No surface space-charge region.)

In the case of real surfaces, an additional complication may enter into the description of the surface recombination process. If, for example, negatively charged ions are present on the surface of the n-type semiconductor sample, as shown in Fig. 5.15, the carrier distribution within the sample will be disturbed by these ions: electrons will be pushed away from the surface and holes will be attracted there. Consequently, space-charge neutrality will no longer hold in a region of thickness x_d near the surface. This region is called a *surface space-charge region*.†

† Surface space-charge regions are discussed in detail in Chapters 9 and 10.

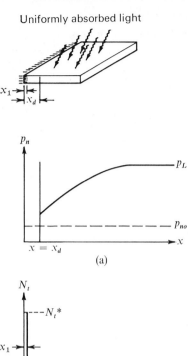

Fig. 5.15 Enhanced recombination rate near a surface due to a high center density, and the resulting distribution of excess minority carriers. (Surface space-charge region induced by negative ions.)

If we now shine light on the sample and generate electron-hole pairs uniformly throughout its interior, the carriers will move to the surface to recombine there as in the previous case. However, because of the presence of the surface space-charge region, additional care must be taken in establishing the balance between the flux of minority carriers to the surface and the surface recombination rate.

The total number of carriers recombining at the surface per unit area and unit time will be

$$U_s = \frac{\sigma_p \sigma_n v_{th} N_{st}[p_s n_s - n_i^2]}{\sigma_n[n_s + n_i e^{(E_t - E_i)/kT}] + \sigma_p[p_s + n_i e^{(E_i - E_t)/kT}]} \quad (5.49)$$

by analogy to Equation 5.40, where n_s and p_s denote the electron and hole concentrations *at the surface*.[2] If for simplicity we assume that the centers

Surface Recombination

are located at energy $E_t = E_i$ (such centers will, in fact, be the most effective), and that the capture cross sections are equal, i.e., $\sigma_p = \sigma_n = \sigma$, U_s becomes

$$U_s = \sigma v_{th} N_{st} \frac{p_s n_s - n_i^2}{n_s + p_s + 2n_i} = s_o \frac{p_s n_s - n_i^2}{n_s + p_s + 2n_i} \tag{5.50}$$

where $s_o \equiv \sigma v_{th} N_{st}$ is *the surface recombination velocity of a surface without a surface space-charge region*, as given by Equation 5.48.

The flux of minority carriers reaching the surface must equal U_s. If the recombination rate within the surface space-charge region is not too high, this flux can be approximated by the flux of minority carriers reaching the edge of the surface space-charge region. Thus,

$$D_p \left.\frac{\partial p_n}{\partial x}\right|_{x=x_d} = s_o \frac{p_s n_s - n_i^2}{n_s + p_s + 2n_i}. \tag{5.51}$$

The above relationship is really a boundary condition for the diffusion problem which describes the distribution of minority carriers within the bulk of the semiconductor sample. A boundary condition is useful only if it is expressed in terms of concentrations and concentration gradients at the boundary, i.e., at the plane $x = x_d$ in this case. Thus we must express the right-hand side of Equation 5.51 in terms of concentrations at the edge of the surface space-charge region rather than at the surface.

We can accomplish this by assuming that the product of electron and hole concentrations will be constant throughout the surface space-charge region even though equilibrium does not prevail. This constant will, of course, be different from n_i^2. Such an assumption is consistently employed in the treatment of space-charge regions, and will be discussed in the next chapter. In the present case it leads to

$$p_s n_s = p_n(x_d) n_n(x_d) \doteq p_n(x_d) N_D \tag{5.52}$$

where N_D is the donor concentration in the semiconductor sample. Noting that $p_{no} N_D = n_i^2$, we obtain

$$D_p \left.\frac{\partial p_n}{\partial x}\right|_{x=x_d} = s_o \frac{N_D}{n_s + p_s + 2n_i} [p_n(x_d) - p_{no}], \tag{5.53}$$

or

$$D_p \left.\frac{\partial p_n}{\partial x}\right|_{x=x_d} = s[p_n(x_d) - p_{no}] \tag{5.54}$$

where

$$s \equiv s_o \frac{N_D}{n_s + p_s + 2n_i} \tag{5.55}$$

is the surface recombination velocity. Thus the surface recombination velocity will take on a more complicated form when a surface space-charge region is present; it will depend not only on the density of surface recombination centers but also on the donor concentration and on the amount of surface charge which determines n_s and p_s. As the charge on the surface is varied, n_s and p_s and consequently the surface recombination velocity itself, will vary.

The surface recombination velocity will go through a maximum when $(n_s + p_s)$ is at a minimum. This will happen when they are both relatively close to the intrinsic carrier concentration n_i. The maximum surface recombination velocity will then be given by

$$s_{\max} = s_o \frac{N_D}{4n_i}. \tag{5.56}$$

5.5 ORIGIN OF RECOMBINATION-GENERATION CENTERS

We have now seen how energy levels introduced into the forbidden gap can facilitate the recombination of excess carriers within a semiconductor by acting as "stepping stones" between conduction band and valence band. We have also shown that the theory of the recombination-generation process[1] relates the characteristic constants of the recombination rate to the density of recombination-generation centers. Thus we have seen that the lifetime in low-level injection is given by

$$\tau_p = \frac{1}{\sigma_p v_{th} N_t} \tag{5.57}$$

for the example of an *n*-type semiconductor, where N_t is the concentration of the centers per unit volume, and σ_p is their capture cross section for holes. Similarly, by extending the theory to recombination taking place at a surface, we have seen that the surface recombination velocity in the absence of a surface space-charge region is given by

$$s_p = \sigma_p v_{th} N_{st} \tag{5.58}$$

for an *n*-type semiconductor, where N_{st} is the density of the centers per unit surface area.

In this section we now briefly consider some examples of the physical origin of recombination-generation centers, both in the bulk and at the surface.

a. Impurities

When introduced into a silicon sample, column III and column V impurities lead to energy levels within the forbidden gap. Because these elements are relatively similar to silicon (a column IV element), the energy levels associated with them will be shallow, i.e., they will be close to the valence- and conduction-band edges. Thus column III and column V impurities become acceptors and donors in silicon.

There are other impurities which, in contrast with column III and V elements, introduce energy levels nearer to the middle of the forbidden gap. Such impurities, examples of which are copper in germanium and gold in silicon, will therefore act as efficient recombination-generation centers.

The properties of gold in silicon have been studied extensively.[3] It is known that there are actually two energy levels associated with each gold atom: an acceptor level (which can be either neutral or negatively charged) near the middle of the energy gap, and a donor level (which can be either neutral or positively charged) about 0.2 ev below the middle. The theory of the recombination mechanism taking place through the action of such multi-level centers is more complicated than the theory of a single-level center given in this chapter; however, in low-level injection we may apply the latter to describe gold in silicon as a reasonable approximation by considering only the acceptor level in n-type, and the donor level in p-type silicon.

The lifetime of silicon samples can be varied by the controlled addition of gold. This is done by depositing gold onto the surface of the sample and then heating the sample at a certain temperature. After a period of time t such that $t \gg W_s^2/D$, where W_s is the thickness of the sample and D is the gold diffusivity, gold will be uniformly distributed in the sample in a concentration corresponding to its solid solubility at that temperature. (The diffusivity and solid solubility of gold are shown as a function of temperature in Figures 3.4 and 3.7, respectively.)

The experimentally observed relationship between gold concentration and lifetime in silicon[4] is shown in Figure 5.16. It is evident that the data follow the inverse relationship indicated by Equation 5.57. The constant of proportionality corresponds to a capture cross section of $\sim 5 \times 10^{-15}$ cm². Several other investigators have reported values within an order of magnitude of this one.[3] Also shown in this figure are the respective temperatures at which the silicon sample was saturated with gold.

As the gold concentration in the silicon sample is increased sufficiently to become comparable to the concentration of the donor or acceptor impurity, another effect has to be taken into account. Each of the deep lying energy levels associated with the gold atoms will, in effect, remove one majority carrier from the conduction-band in the case of an *n*-type semiconductor, or from the valence band in the case of a *p*-type semiconductor. This effect is sometimes referred to as *carrier removal*. Thus,

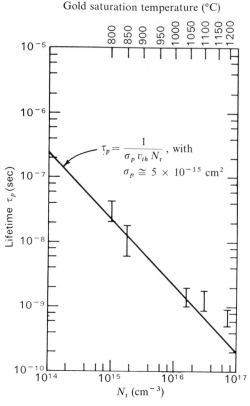

Fig. 5.16 Relationship between lifetime of holes and gold concentration in *n*-type silicon.[4]

for example, in an *n*-type semiconductor the electron concentration will become $n \simeq N_D - N_t$. As a result, the resitivity of the sample will increase with the addition of gold. A similar phenomenon takes place in the case of a *p*-type semiconductor.†

† This simple description loses its validity once the concentration of deep-lying levels approaches the donor or acceptor concentration. Exact calculations of the resistivity of silicon as a function of gold concentration were performed by Boltaks et al.[5] Their results are reproduced also in Bullis's paper.[3]

b. Radiation Damage

Another manner in which energy levels can be introduced into the forbidden gap is by exposure to high-energy radiation—electrons, protons, gamma rays, or neutrons. The high-energy particles can displace atoms from their normal positions in the semiconductor lattice, resulting first in the formation of a vacancy and an interstitial atom. These, in turn,

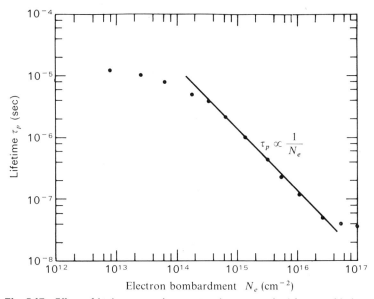

Fig. 5.17 Effect of high energy electron irradiation on the lifetime of holes in n-type silicon.[6]

will rapidly form more complex lattice defects which behave much like impurities introduced into the semiconductor; they will have energy levels within the forbidden gap and can act as acceptors, donors, and recombination centers.

The introduction of recombination centers by high-energy electron irradiation[6] is illustrated by the experimental data shown in Figure 5.17. Here the lifetime of holes in an n-type silicon sample is shown as a function of the total number of bombarding electrons hitting a unit area of the surface of the sample, N_e. It is evident that as bombardment proceeds, the lifetime begins to decrease. If we assume that the incident electrons create recombination centers uniformly within the semiconductor, the total concentration of recombination centers N_t will be given by

$$N_t = N_{to} + KN_e \tag{5.59}$$

where N_{to} is the concentration of recombination-generation centers before irradiation, and K is the probability that an incident electron will create a new recombination center. Making use of Equation 5.57, we then obtain

$$\tau_p = \frac{\tau_{po}}{1 + KN_e/N_{to}} \quad (5.60)$$

which predicts that the lifetime should decrease from its initial value τ_{po} in an inversely proportional manner with increasing radiation dose. This relationship is borne out by the experimental observations shown in Figure 5.17.

The radiation-induced centers will bring about a decrease in the majority carrier concentration (carrier removal) much the same way as gold does in silicon. Thus the resistivity of semiconductors will increase with increasing radiation dose.

At the present time irradiation cannot be readily employed to control the lifetime of semiconductor devices as is possible with impurities like gold. This is because the radiation-induced centers have a tendency to anneal out after a period of time, even at a relatively low temperature. However, the influence of radiation on the lifetime and resistivity of semiconductors is important because many types of semiconductor devices are intended for operation in nuclear reactor or space environments.

c. Surface States

We have seen that a foreign element or even a lattice defect within the semiconductor crystal can introduce energy levels into the forbidden gap. It might be expected that the drastic irregularity associated with a surface of the crystal, where the semiconductor lattice is altogether terminated, should also result in the introduction of a large density of levels into the forbidden gap. Such levels, the existence of which has been predicted theoretically by Tamm and by Shockley,[7] are called *surface states*. If some of these surface states should have energy levels near the middle of the forbidden gap, we would expect them to act as efficient surface recombination centers.

Theoretical estimates of the density of surface states yield values of the same order as the density of surface atoms, $\sim 10^{15}$ cm^{-2}. Such densities have indeed been observed on very clean semiconductor surfaces obtained by cleaving samples under high vacuum. However, germanium and silicon samples after exposure to air for only a few minutes show surface state densities of the order of only $\sim 10^{11}$ cm^{-2}, and thermally oxidized silicon surfaces can show densities yet another order of magnitude smaller.†

---

† Surface states are discussed in further detail in Chapters 9 to 12.

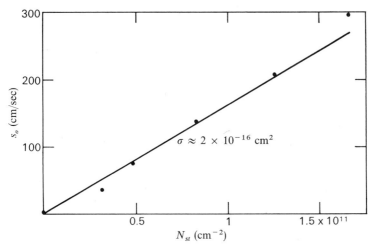

Fig. 5.18 Comparison between surface recombination velocity s_o and surface-state density N_{st}.[8]

The cause of such a reduction is unknown at present; it is evidently brought about by the presence of even a very thin oxide layer such as forms on both germanium and silicon surfaces upon exposure to room air.

Surface recombination velocity s_o values observed on germanium and silicon surfaces are of the order of 10^2 cm/sec; on thermally oxidized silicon, 1 to 10 cm/sec.† Upon irradiation, the magnitude of s_0 as well as the density of surface states has been found to increase. Figure 5.18 shows a comparison between measured values of s_o and independently estimated values of the surface state density N_{st} of thermally oxidized silicon samples.[8] The straight-line relationship observed is in agreement with Equation 5.58, with the slope corresponding to a capture cross section of the same order as that of bulk recombination-generation centers.

READING REFERENCES

A rigorous and thorough treatment of the transport of excess carriers is given by A. K. Jonscher, *Principles of Semiconductor Device Operation*, Wiley, 1960, Chapters 2 and 3.

† Actual measured values of the surface recombination velocity s will not necessarily be the same as s_0, but will be related to it by Equation 5.55. Thus, depending on the particular semiconductor and the temperature (through n_i), on the impurity concentration in the semiconductor, N_D, and on the surface charge density (through $n_s + p_s$), s may be smaller, larger, or the same as s_o.

Band-to-band recombination and recombination through multiple-level centers are discussed in further detail in Chapter 6, J. L. Moll, *Physics of Semiconductors*, McGraw-Hill Book Co., 1964.

A review of and extensive bibliography on "Recombination in Semiconductors" is given by G. Bemski, *Proc. IRE*, **46**, 990 (1958).

Surface recombination is discussed in further detail by A. Many, Y. Goldstein, and N. B. Grover, *Semiconductor Surfaces*, Wiley, 1965, Chapters 5, 7, and 9.

REFERENCES CITED

1. R. N. Hall, "Electron-Hole Recombination in Germanium," *Phys. Rev.*, **87**, 387 (1952); W. Shockley and W. T. Read, "Statistics of the Recombination of Holes and Electrons," *Phys. Rev.*, **87**, 835 (1952).

2. D. T. Stevenson and R. J. Keyes, "Measurements of the Recombination Velocity at Germanium Surfaces," *Physica*, **20**, 1041 (1954).

3. A review of the "Properties of Gold in Silicon" is given by W. M. Bullis, *Solid-State Electron.*, **9**, 143 (1966).

4. A. E. Bakanowski and J. H. Forster, "Electrical Properties of Gold-Doped Diffused Silicon Computer Diodes," *Bell System Tech. J.*, **39**, 87 (1960).

5. B. I. Boltaks, G. S. Kulikov, and R. Sh. Malkovich, "The Effect of Gold on the Electrical Properties of Silicon," *Soviet Physics—Solid State*, **2**, 167 (1960).

6. G. K. Wertheim, "Energy Levels in Electron Bombarded Silicon," *Phys. Rev.*, **105**, 1730 (1957).

7. See, for instance, A. Many, Y. Goldstein, and N. B. Grover, *Semiconductor Surfaces*, Wiley, 1965, Chapters 1 and 5.

8. D. J. Fitzgerald and A. S. Grove, "Radiation-Induced Increase in Surface Recombination Velocity of Thermally Oxidized Silicon Structures," *Proc. IEEE*, **54**, 1601 (1966).

PROBLEMS

5.1 Calculate the resistivity of an n-type silicon sample under illumination. The sample has a lifetime of 1 μsec and its resistivity in dark is 10 Ω cm. The light is absorbed uniformly in the semiconductor, leading to the creation of 10^{22} electron-hole pairs/(cm³ sec). What part of the conductivity is due to:
(a) Majority carriers?
(b) Minority carriers?

5.2 Verify that Equation 5.17 satisfies the appropriate differential equation and boundary conditions.

5.3 A sample of 1 Ω cm n-type silicon is illuminated. The uniformly absorbed light creates 10^{17} electron-hole pairs /(cm³ sec). The lifetime is 10 μsec; the surface recombination velocity 100 cm/sec. Calculate:

Problems

    (a) The number of holes recombining at the surface, per unit surface area, in unit time, and

    (b) The number of holes recombining in the bulk within 3 diffusion lengths of the surface, per unit surface area, in unit time.

5.4 (a) Estimate the electric field near the surface for the above problem.

    (b) Estimate the percentage contribution of the drift-term to the flux of holes to the surface.

5.5 A wafer of silicon doped with 2×10^{16} donor atoms/cm^3 has been saturated with gold at 920°C. It then was oxidized and treated in such a way that it now has 10^{10} surface recombination centers/cm^2.

    (a) Calculate the bulk lifetime, the diffusion length, and the surface recombination velocity in the absence of a surface space-charge region, and also the maximum surface recombination velocity.

    (b) If the sample is illuminated with uniformly absorbed light which creates 10^{17} carrier pairs/(cm^3 sec), what will the hole concentration at the surface and the hole flux to the surface be in the absence of a surface space-charge region?

5.6 What is the electron concentration and the resistivity of the sample in Problem 5.5 before and after gold diffusion?

5.7 Derive the expression for the recombination-generation rate, and indicate if net recombination or generation of carriers takes place, under the following conditions:

    (a) In a region of a semiconductor sample which is completely depleted of carriers (i.e., both n and $p \ll n_i$).

    (b) In a region of a semiconductor sample which is depleted of minority carriers only (for example, $p_n \ll p_{no}$, but $n_n = n_{no}$).

    (c) In a region of a semiconductor in which $n = p = n_o$, where $n_o \gg n_i$.

(These formulas will be used in Chapter 6 in the derivation of the current-voltage characteristics of p-n junctions.)

5.8 In a region of semiconductor which is completely depleted of carriers (i.e., n and $p \ll n_i$) electron-hole pairs are generated by alternate emission of electrons and of holes by the centers. Derive the average time that takes place between such emission processes. (This time is called the *emission time constant* of the centers.) Estimate its value for centers with $E_t = E_i$, in silicon.

5.9 Considering a p-type semiconductor, derive the energy level of those centers from which a trapped electron is as likely to be re-emitted into the conduction band as it is to recombine with holes. Will centers with energy above or below this level be efficient recombination centers?

5.10 Consider recombination-generation centers uniformly distributed in energy throughout the forbidden gap, with density D_t (cm^{-2} ev^{-1}).

    (a) By superposition of the effect of such centers, and assuming $\sigma_p = \sigma_n = \sigma$, derive the formula for the net recombination-generation rate U.

    (b) Derive the expression for the generation rate in a region which is completely depleted of carriers (i.e., n and $p \ll n_i$).

    (c) Derive the expression for the lifetime in low-level injection.

Compare each of the above formulas with the corresponding ones for single level centers located at $E_t = E_i$, and discuss the similarities and differences.

TABLE 5.1

IMPORTANT FORMULAS FOR SEMICONDUCTORS
UNDER NON-EQUILIBRIUM CONDITIONS
Midgap recombination-generation centers, i.e., $E_t = E_i$
Equal capture cross-sections, i.e., $\sigma_p = \sigma_n = \sigma$

	n-Type semiconductor	p-Type semiconductor
Net bulk recombination rate per unit volume	$U = \dfrac{1}{\tau}(p_n - p_{no})$	$U = \dfrac{1}{\tau}(n_p - n_{po})$
Net surface recombination rate per unit area	$U_s = s[p_n(0) - p_{no}]$	$U_s = s[n_p(0) - n_{po}]$
Lifetime	$\tau = \dfrac{1}{\sigma v_{th} N_t}$	$\tau = \dfrac{1}{\sigma v_{th} N_t}$
Surface recombination velocity	$s = s_o \dfrac{N_D}{n_s + p_s + 2n_i}$	$s = s_o \dfrac{N_A}{n_s + p_s + 2n_i}$

$$s_o \equiv \sigma v_{th} N_{st}$$

- **ELECTROSTATICS**
- **SPACE-CHARGE REGION FOR:**
  STEP JUNCTIONS
  LINEARLY GRADED JUNCTIONS
  DIFFUSED JUNCTIONS
- **CAPACITANCE-VOLTAGE CHARACTERISTICS**
- **CURRENT-VOLTAGE CHARACTERISTICS**
- **JUNCTION BREAKDOWN**
- **TRANSIENT BEHAVIOR**

6

p-n Junctions

In the preceding two chapters we have discussed the behavior of homogeneous semiconductor samples both under equilibrium and non-equilibrium conditions. Most semiconductor devices, however, incorporate both *p*- and *n*-type regions, and it is the *junction* between these regions that leads to their useful electrical characteristics.

A *planar silicon n^+p junction diode*[†] is illustrated schematically in Figure 6.1a. Such a diode may be fabricated by first growing a relatively lowly doped *p*-type epitaxial film upon a highly doped substrate of the same type; and then thermally oxidizing the surface of the silicon. Windows are then opened in the silicon dioxide layer, and donor impurities are permitted to diffuse into the silicon where the oxide layer had been removed.[‡] This results in the formation of the n^+p junction illustrated in the figure. (The superscripts $+$ or $-$ indicate regions of relatively high or low impurity concentrations, respectively.) Many such junctions are formed simultaneously on a wafer. After they are cut apart by scribing, contacts are attached to both the bottom side and to the diffused portion of each diode, and the diodes are inserted into a package.

[†] The term *junction* denotes the boundary between the two semiconductor regions, whereas the term *diode* refers to the finished semiconductor device incorporating a single junction. These terms sometimes are used interchangeably.

[‡] Epitaxial growth, thermal oxidation, and solid-state diffusion are discussed in detail in Part I.

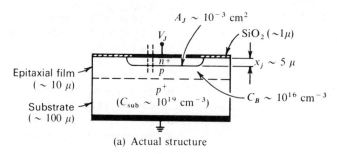

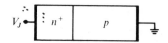

(b) Idealized one-dimensional model

Fig. 6.1 A typical planar n^+p junction diode. Representative impurity concentrations and dimensions are indicated.

For simplicity, we study an idealized one-dimensional model of this n^+p junction, which is illustrated in Figure 6.1b. This model can be considered as a section of the planar diode along the dotted lines shown in Figure 6.1a. This idealized model also neglects the variation of the impurity concentration in the p-region due to the epitaxial construction. Where this concentration variation or the characteristic shape of the planar junction leads to deviations, we modify our discussion accordingly.

The most important characteristic of p-n junctions is that they *rectify*, i.e., they permit the passage of electric current in only one direction. This is illustrated in Figure 6.2 where we show the current-voltage characteristic of a typical silicon p-n junction diode whose structural parameters are approximately as indicated in Figure 6.1. It is evident that when a negative voltage is applied to the n-region, a current begins to flow at a very small applied voltage. In contrast, when a positive voltage is applied to the n-region, no current flow is perceived at the scale of this illustration. Eventually, when a large enough positive voltage is applied to the n-region, current does begin to flow again. This condition is referred to as *junction breakdown*. The bias condition corresponding to easy conduction is called *forward bias*, and the bias condition corresponding to no conduction is called *reverse bias*. The current-voltage characteristic under forward bias is also shown with an expanded voltage scale in the lower half of Figure 6.2.

Diodes themselves are very important and useful because of their special current-voltage characteristics; for instance, they can be employed

p-n Junctions

as rectifiers, or as switches in digital computers. However, the importance of *p-n* junctions is broader than the use of diodes alone would indicate. Two of the most important semiconductor devices, junction transistors and junction field-effect transistors (which form the subject of the next

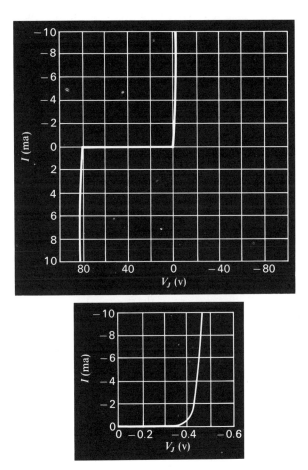

Fig. 6.2 The current-voltage characteristics of a silicon n^+p diode. The structural parameters of this diode are approximately as indicated in Fig. 6.1.

two chapters), consist of two *p-n* junctions in the vicinity of each other. Their characteristics will be seen to be a direct consequence of the characteristics of single *p-n* junctions. In addition, in Part III we will see that the analysis of surface phenomena and of surface-controlled devices can be greatly simplified by relating it to the analysis of *p-n* junctions.

Accordingly, in this chapter we consider *p-n* junctions in detail. We begin with a brief review of the relevant principles of electrostatics, and then apply these principles to the study of the space-charge regions of diffused junctions and their two limiting forms: step junctions and linearly graded junctions. We then consider the capacitance-voltage and the current-voltage characteristics of *p-n* junctions, and junction breakdown. Finally, we give a brief discussion of transient phenomena.

6.1 ELECTROSTATICS

a. Connection between Band Structure and Electrostatic Quantities

The *electric field* $\mathscr{E}$ is defined as the force acting on a unit positive charge. Accordingly, the force acting on an electron, which has a charge $-q$, is $-q\mathscr{E}$.

In general, we know that a force is equal to the negative of the gradient of potential energy. Accordingly, the force acting on an electron equals

$$-q\mathscr{E} = -(\text{gradient of potential energy of electrons}).$$

We saw in Chapter 4 that the potential energy of an electron is represented by the lower edge of the conduction band, E_c. However, since we are interested only in the gradient of the potential energy, we can also use any part of the energy bands that is displaced from the conduction-band edge by a fixed amount. Thus we could equally well choose the gradient of the valence-band edge E_v or of the intrinsic Fermi level E_i in the above expression rather than the gradient of E_c. In practice it is frequently convenient to use E_i. Accordingly, we have the expression

$$\mathscr{E} = \frac{1}{q} \text{grad } E_i \tag{6.1}$$

or, in one dimension,

$$\mathscr{E} = \frac{1}{q} \frac{dE_i}{dx}. \tag{6.2}$$

The *electrostatic potential* ϕ is the quantity whose gradient is the negative of the electric field, i.e., it is defined by the equation

$$\mathscr{E} = -\text{grad } \phi \tag{6.3}$$

or, in one dimension,

$$\mathscr{E} = -\frac{d\phi}{dx}. \tag{6.4}$$

Space-Charge Region for Step Junctions

Comparison of Equation 6.3 with 6.1 yields

$$\phi = -\frac{E_i}{q} \tag{6.5}$$

which provides the relationship between the electrostatic potential and the potential energy of an electron.

b. Poisson's Equation

Poisson's equation states that

$$\frac{d^2\phi}{dx^2} = -\frac{\rho}{K\epsilon_0} \tag{6.6}$$

or

$$\frac{d^2 E_i}{dx^2} = \frac{q\rho}{K\epsilon_0} \tag{6.7}$$

or

$$\frac{d\mathscr{E}}{dx} = \frac{\rho}{K\epsilon_0} \tag{6.8}$$

where ρ is the charge density per unit volume, conveniently expressed in semiconductor work in units of e/cm^3 where e is the magnitude of the electronic charge, 1.60×10^{-19} coulombs; K is the dielectric constant, and ϵ_0 is the permittivity of free space, 8.86×10^{-14} f/cm = 55.4 e/vμ in units convenient in semiconductor work.

It is evident from Equation 6.8 that the electric field is obtained by integrating the charge distribution as a function of distance. Simple examples of such integrations are illustrated in Figure 6.3.

In Figure 6.3a we show a charge distribution given by a delta function containing a total charge per unit cross-sectional area of Q. Integration of this charge distribution results in a step change in electric field by the amount $Q/K\epsilon_0$, as shown in the figure. In Figure 6.3b we show a uniform charge distribution over the distance L. In this case the electric field increases linearly with distance, but the total increase in the electric field is again related only to the total charge contained in the region, $\rho_0 L$.

6.2 SPACE-CHARGE REGION FOR STEP JUNCTIONS

Many important characteristics of *p-n* junctions are associated with a *space-charge region* (a region where $\rho \neq 0$), formed between the *n*-type and *p*-type regions. Accordingly, we now consider what happens when an

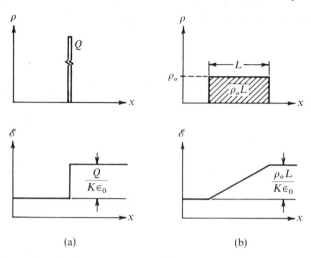

Fig. 6.3 Illustration of the consequences of Poisson's equation for two simple one-dimensional cases.

n-type semiconductor and a p-type semiconductor are brought into intimate contact. Bringing separate pieces of semiconductor into intimate contact is not a practical way of making useful p-n junction devices but we employ this scheme to illustrate the principles involved.

a. Equilibrium Case

First we consider the case with no bias applied to either semiconductor region. The n-type and p-type samples before contact are illustrated in Figure 6.4a. As we discussed in Chapter 4, an n-type semiconductor is characterized primarily by the fact that it contains a large concentration of electrons. Similarly, a p-type semiconductor sample is characterized by the fact that it, in turn, contains a large concentration of holes. If these two samples are brought into intimate contact, as illustrated in Figure 6.4b, a flux of electrons and of holes will flow in such directions as to even out the large concentration gradients existing between the two regions.

If the electrons and holes were not charged, these fluxes would continue until a uniform concentration of both species was established throughout the combined semiconductor sample. Because they are charged, and because the semiconductor samples also contain ionized impurity ions, the situation will be entirely different, as illustrated in Figure 6.4b. When a net flow of electrons from the n-region to the p-region, and of holes from the p-region to the n-region takes place, a space charge due to donor and acceptor ions is formed. Hence an electric field results in the vicinity of

Space-Charge Region for Step Junctions

the junction as indicated by the arrows in Figure 6.4b. This electric field is of such direction as to keep the holes in the *p*-region and the electrons in the *n*-region. Thus, *after a certain number of electrons and holes have flowed from one region to the other, an electric field will be built up, preventing further net flow of the carriers.*

This is the manner in which equilibrium is reached. To restate, in equilibrium the *net* flux of both holes and of electrons will be zero; *the*

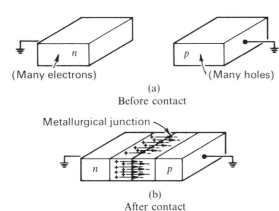

Fig. 6.4 Schematic illustration of a *p-n* junction in equilibrium.
(a) *n*- and *p*-type semiconductors.
(b) Junction between the two.

diffusion flux of each carrier at the p-n junction will be precisely equal and opposite to the flux of that carrier due to the electric field.

Let us now use this criterion to obtain the condition for equilibrium in quantitative terms. The net flux of holes F_p is given by

$$F_p = -D_p \frac{dp}{dx} + \mu_p \mathscr{E} p. \tag{6.9}$$

Substituting the expression for the hole concentration,

$$p = n_i e^{(E_i - E_F)/kT} \tag{6.10}$$

and its derivative,

$$\frac{dp}{dx} = \frac{p}{kT}\left[\frac{dE_i}{dx} - \frac{dE_F}{dx}\right], \tag{6.11}$$

and incorporating Einstein's relationship $\mu_p = qD_p/kT$, the net flux of holes is then given by

$$F_p = \frac{D_p}{kT} p \frac{dE_F}{dx} = \frac{1}{q} \mu_p p \frac{dE_F}{dx}. \tag{6.12}$$

Thus the condition of zero net hole flux means that the Fermi level must be uniform as we pass from the *n*-region to the *p*-region. Similar considerations applied to the electron flux lead to the formula

$$F_n = -\frac{D_n}{kT} n \frac{dE_F}{dx} = -\frac{1}{q}\mu_n n \frac{dE_F}{dx} \tag{6.13}$$

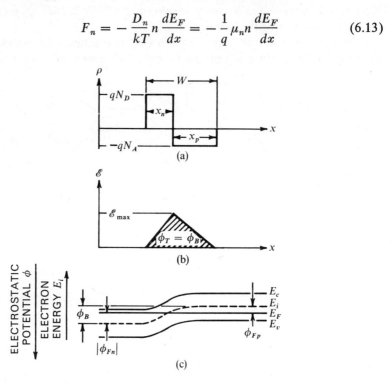

Fig. 6.5 Schematic illustration of the (a) charge, (b) electric field, and (c) potential distribution within a step junction in equilibrium (depletion approximation).

according to which the condition of zero net electron flux leads to the same requirement. Thus *in equilibrium the Fermi level must be constant throughout the semiconductor.*†

To calculate the characteristics of the space-charge region formed near the junction, we assume that it consists entirely of the charge of the ionized acceptors and donors. This assumption is equivalent to saying that most of the space-charge region will be completely depleted of carriers or, in other words, that the carrier concentrations *n* and *p* will be negligibly

---

† As mentioned in Chapter 4, the Fermi level can be considered as the *chemical potential* of electrons. Since the condition for equilibrium in any system is that the chemical potential should be constant as we pass from one part of the system to the other, the constancy of the Fermi level in equilibrium follows.

Space-Charge Region for Step Junctions

small in comparison to the impurity concentration over most of the space-charge region. This is called the *depletion approximation*.†

The charge distribution in the semiconductor sample as a function of distance, subject to this simplifying assumption, is illustrated in Figure 6.5a for the case of uniform impurity concentration in both *p* and *n* regions. Such a junction is called an abrupt or *step junction*. The density of space charge on the *n* and *p* sides of the metallurgical junction will be given by qN_D and by $-qN_A$, respectively. The widths of the respective portions of the space-charge region are designated by x_n and x_p, with the total width $(x_n + x_p)$ being denoted by W. The field distribution, obtained by integration of the charge distribution, is illustrated in Figure 6.5b. Integration of the electric field distribution in turn leads to the electron energy E_i as a function of distance as illustrated in Figure 6.5c. The conduction- and valence-band edges, of course, follow the variation of the intrinsic Fermi level.

Let us now consider what determines the total amount of bending of the bands, i.e., the total amount of variation of either electron energy or electrostatic potential, as we cross from the *p*-region to the *n*-region. It is evident from Figure 6.5c that, since the Fermi level is constant, this variation will consist of the sum of the absolute values of the two *Fermi potentials* ϕ_{Fp} and ϕ_{Fn} which are defined by

$$\phi_{Fp} \equiv -\left.\frac{E_F - E_i}{q}\right|_{p\text{-region}}, \qquad \phi_{Fn} \equiv -\left.\frac{E_F - E_i}{q}\right|_{n\text{-region}} \qquad (6.14)$$

Thus ϕ_{Fp} is positive while ϕ_{Fn} is negative. (Hence the need to take the absolute value of the latter.)

To calculate the Fermi potentials, we consider that in the neutral *p*-region (i.e., outside the space-charge region) $p \doteq N_A$, while in the neutral *n*-region $n \doteq N_D$. Then, from Equation 6.10 we get

$$\phi_{Fp} = \frac{kT}{q} \ln \frac{N_A}{n_i}. \qquad (6.15)$$

Likewise,

$$\phi_{Fn} = -\frac{kT}{q} \ln \frac{N_D}{n_i}. \qquad (6.16)$$

The total potential variation ϕ_T in equilibrium then will be the sum of the absolute values of the Fermi potentials,

$$\phi_T = \phi_{Fp} + |\phi_{Fn}|. \qquad (6.17)$$

This quantity is often referred to as the *built-in voltage* ϕ_B of a *p-n* junction.

† In this context, the terms *space-charge region* and *depletion region* are used interchangeably.

The magnitude of the Fermi potential on either side of a junction is given as a function of the magnitude of the net impurity concentration $C_B \equiv |N_D - N_A|$ in Figure 6.6 for silicon at room temperature, based on Equations 6.15 or 6.16. This enables the calculation of ϕ_B for any given step junction.

Because the electric field in the neutral regions of the semiconductor must be zero, we can immediately see that the total charge per unit area

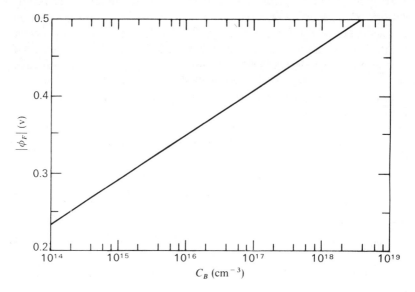

Fig. 6.6 Fermi potential versus net impurity concentration (Silicon, 300°K). $\phi_F > 0$ for p-type region; $\phi_F < 0$ for n-type region.

on either side of the p-n junction must be precisely equal and opposite. Thus,

$$N_D x_n = N_A x_p. \tag{6.18}$$

The maximum electric field in the p-n junction is then

$$\mathscr{E}_{max} = \frac{qN_D x_n}{K_s \epsilon_0} = \frac{qN_A x_p}{K_s \epsilon_0}. \tag{6.19}$$

Inspection of Figure 6.5b shows that the total potential variation—the area under the field triangle—is given by

$$\phi_T = \tfrac{1}{2} \mathscr{E}_{max} W. \tag{6.20}$$

We can now readily obtain the total depletion region width W of a step junction as a function of the total electrostatic potential variation from one

Space-Charge Region for Step Junctions

side of the junction to the other ϕ_T,

$$W = \sqrt{\frac{2K_s\epsilon_0}{q} \frac{N_A + N_D}{N_A N_D}} \phi_T. \tag{6.21}$$

A particularly important case of step junctions which is often encountered in practice is a step junction where the concentration of the impurity on one side of the junction is much larger than that on the other side of the junction, for example, $N_D \gg N_A$. This results in the simpler form of Equation 6.21,

$$W = \sqrt{\frac{2K_s\epsilon_0}{qN_A}} \phi_T. \tag{6.22}$$

An alloy junction and a very "shallow" diffused junction are both very much like *one-sided step junctions*. (We put the word "shallow" in quotation marks because what is "shallow" and what is "deep" depends on a number of factors as we shall see later.) The concentration distribution for an example of a shallow diffused n^+p junction is shown in Figure 6.7a. The calculations shown in this figure are for a complementary error-function type impurity distribution with a relatively high surface concentration, $C_S = 10^{20}$ cm^{-3}; low bulk concentration, $C_B = 10^{14}$ cm^{-3}, and a relatively shallow junction depth of $x_j = 1$ μ.

The charge distribution due to such a concentration distribution as well as the corresponding one-sided step-junction approximation are shown in Figure 6.7b, both assuming complete depletion of carriers. The corresponding electric field and potential distributions are shown in Figure 6.7c and d.

These distributions were calculated as follows. First, Poisson's equation was integrated to obtain the electric field distribution. This yields

$$\mathscr{E}(x) = \mathscr{E}(0) - q\frac{C_B x}{K_s\epsilon_0} \tag{6.23}$$

where the origin, $x = 0$, is taken to be at the junction.

Applying the boundary condition $\mathscr{E}(W) = 0$ yields

$$\mathscr{E}(0) = \mathscr{E}_{\max} = \frac{qC_B W}{K_s\epsilon_0} \tag{6.24}$$

and

$$\mathscr{E}(x) = \mathscr{E}_{\max}\left(1 - \frac{x}{W}\right) \tag{6.25}$$

which is shown in Figure 6.7c.

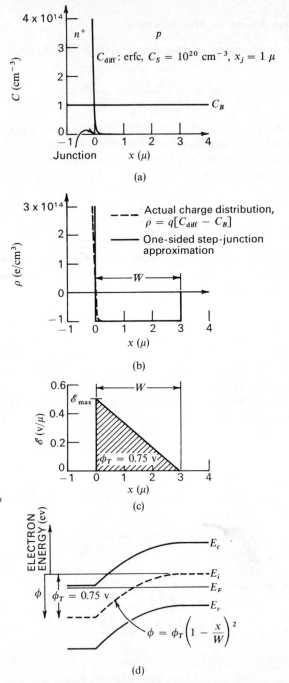

Fig. 6.7 The one-sided step-function approximation. (a) Concentration distribution for a "shallow" diffusion. (b) Charge distribution. (c) Electric field distribution. (d) Potential distribution. Calculations are for silicon at 300°K.

Space-Charge Region for Step Junctions

Integrating once again to obtain the distribution of electron energy yields

$$E_i(x) = q \, \mathscr{E}_{\max}\left(x - \frac{x^2}{2W}\right) + \text{constant.} \tag{6.26}$$

Taking as an arbitrary choice of zero electron energy the electron energy in the neutral p-region or $E_i(W) = 0$ and employing Equation 6.20 yields

$$E_i(x) = -q\phi_T\left(1 - \frac{x}{W}\right)^2 \tag{6.27}$$

or

$$\phi(x) = \phi_T\left(1 - \frac{x}{W}\right)^2 \tag{6.28}$$

where

$$\phi_T = \phi_{Fp} + |\phi_{Fn}| = \frac{qC_B W^2}{2K_s\epsilon_0}. \tag{6.29}$$

This distribution is shown in Figure 6.7d.†

Important relationships pertaining to one-sided step junctions are summarized in Table 6.1 at the end of this chapter.

b. Reverse Bias and Forward Bias

If we apply a positive voltage V_R to the n-region with respect to the p-region, the total electrostatic potential variation across the junction ϕ_T will increase by that amount. Thus,

$$\phi_T = \phi_B + V_R. \tag{6.30}$$

As a result, the width of the space-charge region on both the n-side and on the p-side will increase. This situation is illustrated in Figure 6.8 where

---

† It is difficult to actually calculate the Fermi potential on the heavily doped side of such a junction. In practice, we can assume that the Fermi level will just touch the appropriate band edge so that the Fermi potential on the heavily doped side will be just half of the energy gap, or 0.55 ev for silicon. An alternative assumption, which often leads to results in closer agreement with experiment, is to take $\phi_B \cong 2\phi_F$ where ϕ_F is the Fermi potential on the lowly doped side.

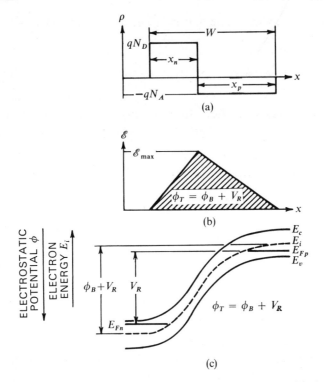

Fig. 6.8 Schematic illustration of the (a) charge distribution, (b) electric field, and (c) potential distribution within a reverse-biased step-junction. (Depletion approximation.)

we indicate the increase in both the maximum electric field and the total potential variation from one side of the junction to the other as well as the widening of the space-charge region.

Substitution of Equation 6.30 into 6.21 or 6.22 yields the width of the space-charge region as a function of applied reverse voltage V_R. Calculations are shown in Figure 6.9 for the one-sided step-junction approximation with substrate impurity concentration as parameter. The depletion region width is relatively constant while the reverse bias V_R is small in comparison to the built-in voltage ϕ_B. When the applied reverse voltage becomes large in comparison to the built-in voltage, the depletion region width increases with the square root of the applied reverse voltage.

The *quasi-Fermi levels* associated with the reverse-biased *p-n* junction, E_{Fn} and E_{Fp}, are also indicated in Figure 6.8. Their significance will be discussed in relation to forward currents in Section 6.6.

The forward-bias condition, in which a negative voltage V_F is applied to the *n*-region relative to the *p*-region, can be considered in a similar

Space-Charge Region for Linearly Graded Junctions

manner. In this case V_F is negative and therefore the total variation of electrostatic potential across the junction ϕ_T is smaller than in equilibrium, bringing about a narrowing of the space-charge region. However, as we shall see in a later section, in this condition large currents—many orders of magnitude larger than in the reverse-bias case—flow across the *p-n* junction. Corresponding to these large currents, the carrier concentrations

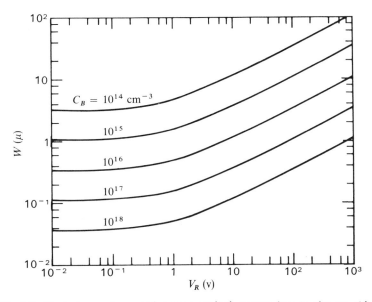

Fig. 6.9 Depletion region width versus applied reverse bias in the one-sided step-junction approximation. (Silicon, 300°K.)

in the space-charge region will become less and less negligible in comparison to the impurity ion concentration so that the depletion approximation loses its validity.

6.3 SPACE-CHARGE REGION FOR LINEARLY GRADED JUNCTIONS

Whereas the one-sided step-junction approximation provides an adequate description for alloy junctions and for "shallow" diffused junctions, in the case of "deep" diffused junctions we must use another simple approximation, called the *linearly graded junction* approximation. In Figure 6.10 a complementary error-function distribution is shown with the same surface concentration and the same substrate concentration as the one in Figure 6.7. The only difference is that the diffusion in this case is continued long enough to make the junction depth 100 μ instead of the 1 μ as in Figure 6.7.†

While it is evident that the one-sided step-junction approximation could not be applied to describe such a distribution, it also appears from Fig. 6.10 that the net impurity distribution, or the charge distribution, can be well approximated by a straight line throughout the depletion region. This straight-line charge distribution, when integrated by using Poisson's

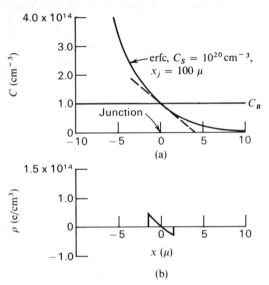

Fig. 6.10 The linearly graded junction approximation.
(a) Concentration distribution for "deep" diffusion.
(b) Charge distribution.

equation, leads to a parabolic electric field distribution and a cubic potential distribution. In particular, the width of the depletion region will be given by

$$W = \left[\frac{12K_s\epsilon_0\phi_T}{qa}\right]^{1/3}, \quad (6.31)$$

the maximum electric field by

$$\mathscr{E}_{max} = 1.5\frac{\phi_T}{W}, \quad (6.32)$$

and the built-in voltage by

$$\phi_B = \frac{2kT}{q}\ln\frac{aW_0}{2n_i} \quad (6.33)$$

† This is an extremely deep diffusion used only to emphasize the difference between the two cases.

Space-Charge Region for Linearly Graded Junctions

where W_0 is the width of the zero-bias (equilibrium) depletion region, and

$$a \equiv \frac{dC}{dx}\bigg|_{x=x_j}$$

is the *impurity concentration gradient at the junction*—the principal characteristic of linearly graded junctions.

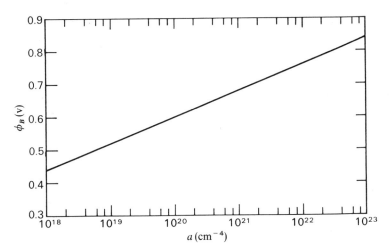

Fig. 6.11 Built-in voltage for linearly graded junctions. (Silicon, 300°K.)

In Figure 6.11 the built-in voltage of linearly graded junctions in silicon at room temperature is shown as a function of a. This figure was obtained by solving the transcendental equation that results when W_o is eliminated between Equations 6.31 and 6.33. Figure 6.12 shows the depletion region width as a function of applied reverse bias V_R for 1-μ and 10-μ deep diffused junctions, based on the linearly graded junction approximation.

It should not be inferred from these figures that the linearly graded junction approximation is actually valid throughout the range of these calculations. For instance, it can be seen from Figure 6.10 that, as the depletion region with W increases, the straight-line approximation begins to deviate increasingly from the actual charge distribution. The linearly graded junction approximation becomes rather meaningless for

$$W > \frac{C_B}{a/2}. \tag{6.34}$$

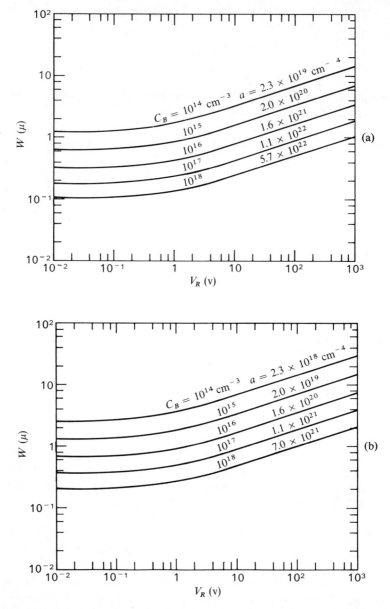

Fig. 6.12 Depletion region width versus applied reverse bias in the linearly graded junction approximation. Erfc distribution, $C_S = 10^{20}$ cm^{-3}.
(a) $x_j = 1\ \mu$.
(b) $x_j = 10\ \mu$.
(Silicon, 300°K.)

6.4 SPACE-CHARGE REGION FOR DIFFUSED JUNCTIONS

We have considered two limiting cases of diffused junctions: shallow junctions that can be well represented by the one-sided step-junction approximation, and deep junctions which can be well represented by the linearly graded junction approximation. Practical diffused p-n junctions may be approximated by one or the other of these, or they may be in between.

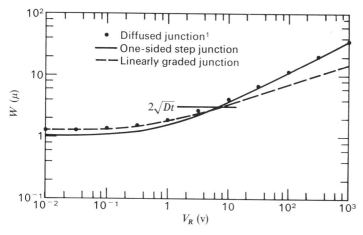

Fig. 6.13 Depletion region width versus applied reverse bias for a diffused junction: comparison with the one-sided step-junction and linearly graded junction approximations. Erfc distribution, $C_S = 10^{20}$ cm^{-3}, $C_B = 10^{15}$ cm^{-3}, $x_j = 10\,\mu$.

Extensive numerical integration of Poisson's equation for diffused junctions has been performed by Lawrence and Warner.[1] We now consider some of their results plotted in a form which enables easy comparison with the two simple closed-form approximations considered above. In Figure 6.13 we show the depletion region width as a function of applied reverse bias for a fairly typical diffused junction (Silicon, room temperature). The points represent the results of Lawrence and Warner,[1] while the lines correspond to the one-sided step-junction and the linearly graded junction approximations, respectively. This figure demonstrates that a given p-n junction cannot by itself be classified as either linearly graded or step; the particular one considered in this figure follows the linearly graded junction approximation at low bias voltages and the one-sided step-junction approximation at high bias voltages.

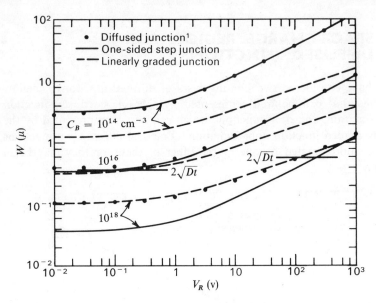

Fig. 6.14 Depletion region width versus applied reverse bias for junctions diffused into substrates of various impurity concentrations: comparison with the one-sided step-junction and linearly graded junction approximations. Erfc distribution, $C_S = 10^{20}$ cm^{-3}, $x_j = 1\ \mu$.

This interesting fact is given a simple explanation if we consider that the relative steepness of the impurity concentration gradient should depend on the size of the depletion region which, in turn, depends on the bias voltage. Thus we have to compare the characteristic length of the diffusion—the diffusion length $2\sqrt{Dt}$—with the characteristic length of the space-charge region W to determine whether a particular junction is in the step or in the linearly graded junction regime. This is further illustrated by the calculations shown in Figures 6.14 and 6.15 where Lawrence and Warner's calculations are compared with the two approximations for various substrate impurity concentrations, for 1- and 10-μ junction depths, respectively. Wherever possible, the diffusion length $2\sqrt{Dt}$ corresponding to a particular junction is indicated.

A useful rule-of-thumb criterion can be inferred from these figures: When the depletion region width W is larger than the diffusion length $2\sqrt{Dt}$ of the junction-forming impurity diffusion, the one-sided step junction approximation is better. Conversely, when the depletion region width is smaller than the diffusion length, the linearly graded junction approximation is preferable.

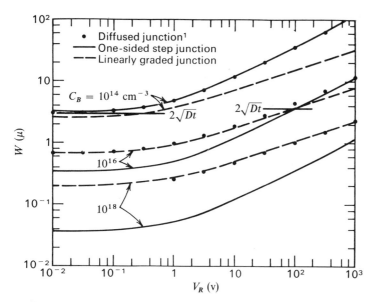

Fig. 6.15 Depletion region width versus applied reverse bias for junctions diffused into substrates of various impurity concentrations: comparison with the one-sided step-junction and linearly graded junction approximations. Erfc distribution, $C_S = 10^{20}$ cm^{-3}, $x_j = 10\ \mu$.

6.5 CAPACITANCE-VOLTAGE CHARACTERISTICS

Capacitance per unit area is defined[†] as $C \equiv dQ/dV$, where dQ is the incremental increase in charge per unit area upon an incremental change in the applied voltage dV.

Let us now calculate the capacitance of a *p-n* junction with an arbitrary impurity distribution, such as shown in Figure 6.16. The charge and electric field distributions designated by the solid line correspond to a voltage V applied to the *n*-region. If this voltage is increased by an amount dV, the charge distribution and the electric field distribution will both change to that indicated by the dashed line. The incremental charge dQ corresponds to the area between the two charge-distribution curves on either side of the depletion region, while the magnitude of the change in the applied voltage dV is indicated by the cross-hatched area between the two electric field distribution curves.

The increase in depletion layer width and the corresponding increase in charge on the *n*-side of the depletion region will bring about an increase in the electric field by an amount $d\mathscr{E} = dQ/K_s\epsilon_0$. (See Section 6.1.) The

[†] This definition yields the *small-signal* capacitance.

corresponding change in the applied voltage dV is approximately $(d\mathscr{E})\,W$ which equals $(dQ/K_s\epsilon_0)W$. Thus, by the definition of the capacitance per unit area, we find that

$$C = \frac{K_s\epsilon_0}{W}. \tag{6.35}$$

This equation holds for any arbitrary impurity distribution.

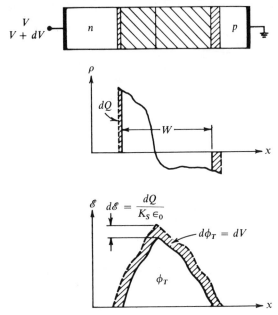

Fig. 6.16 Illustration of the change in charge and electric field distribution resulting from a change in applied reverse bias.

The capacitance per unit area of a parallel-plate capacitor is given by the dielectric constant divided by the separation between the plates. Thus the p-n junction capacitor can be regarded as a parallel-plate capacitor where the distance between the two plates—the distance between the regions where the incremental charge appears—is just the depletion region width W.

In deriving Equation 6.35 we tacitly assumed that all the extra charge that appears in the semiconductor upon the change in applied voltage appears as a change in the charge contained within the depletion region. This is certainly a good assumption in the reverse-bias condition. In the forward-bias condition, however, large currents can flow across the depletion region. Corresponding to these large currents, there will be a considerable charge due to mobile carriers—electrons and holes—present within the depletion region. The rearrangement of these mobile carriers will contribute an additional term to the capacitance.

Because of the very simple relationship between the reverse-bias junction capacitance and the width of the depletion region, measurements of junction capacitance can provide useful information about the nature of the impurity distribution. For example for a one-sided step junction, if we combine Equations 6.22 and 6.35 we obtain

$$C = \sqrt{\frac{qK_s\epsilon_0 C_B}{2(V_R + \phi_B)}} \qquad (6.36)$$

which can be rearranged to yield

$$\frac{1}{C^2} = \frac{2}{qK_s\epsilon_0 C_B}(V_R + \phi_B). \qquad (6.37)$$

It is evident that by plotting $1/C^2$ versus V_R, a straight line should result if the actual impurity distribution can be approximated by the one-sided step-junction theory. In such a case, the slope yields the impurity concentration in the substrate C_B while the intercept yields the built-in voltage ϕ_B.

Similarly, for linearly graded junctions, we can plot $1/C^3$ versus V_R, and obtain a and ϕ_B from slope and intercept if a straight line results.

In general, however, we can evaluate an *arbitrary* impurity distribution with fairly good accuracy from the capacitance-voltage characteristics.[2] This general method is illustrated in Figure 6.17. We consider the simplest case, that of a shallow diffused n^+p junction with an arbitrarily varying impurity distribution on the lower doped side. As before, the charge on the heavily doped side of the depletion region will change by dQ as the applied reverse voltage is changed by dV, where $dQ = qN(W)\,dW$. Correspondingly, the electric field will change by an amount

$$\frac{dQ}{K_s\epsilon_0} = \frac{qN(W)\,dW}{K_s\epsilon_0}.$$

The corresponding change in applied voltage is approximately

$$dV = (d\mathscr{E})\,W = \frac{qN(W)\,d(W^2)}{2K_s\epsilon_0}. \qquad (6.38)$$

Substituting for W^2 from Equation 6.35 leads to an expression for the impurity concentration at the edge of the space-charge region:

$$N(W) = \frac{2}{qK_s\epsilon_0}\frac{1}{d(1/C^2)/dV}. \qquad (6.39)$$

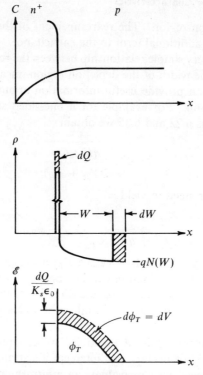

Fig. 6.17 Illustration of the concentration, charge, and electric field distribution leading to the measurement of the impurity distribution.

Thus measurements of the capacitance per unit area C as a function of reverse voltage, with appropriate differentiation, can provide the impurity distribution directly. This method has been used extensively in semiconductor work. Among other places, it was employed in the investigation of impurity redistribution during thermal oxidation and impurity redistribution during epitaxial growth which were discussed in Chapter 3.

6.6 CURRENT-VOLTAGE CHARACTERISTICS[3]

a. Reverse Bias

Under equilibrium conditions electron-hole pairs are generated continually everywhere within the semiconductor. In the absence of an applied voltage, the electron-hole pairs recombine and therefore no current flow results. However, when a positive voltage is applied to the n-region of

Current-Voltage Characteristics

a diode with respect to the *p*-region as shown in Figure 6.18, the electron-hole pairs, once generated, will be separated and their probability of recombination is diminished. This is the basic mechanism giving rise to all types of reverse currents observed in *p-n* junction diodes. All such currents, accordingly, *are due to electron-hole pairs generated someplace in the semiconductor.* Various contributions to the reverse current are

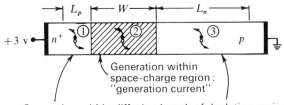

(a)

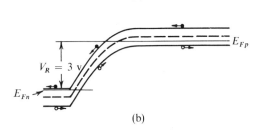

(b)

Fig. 6.18 Illustration of the mechanisms giving rise to reverse current.
(a) n^+p Diode under 3-v reverse bias.
(b) Corresponding band diagram.

distinguished by the region of the semiconductor where an electron-hole pair was generated. Thus, when an electron-hole pair giving rise to the current is generated within the reverse-biased depletion region of the junction, region 2 in Figure 6.18, we refer to the resulting current component as the *generation current*. When electron-hole pairs are generated in the neutral regions 1 and 3 in Figure 6.18 and the carriers diffuse to the reverse-biased junction, thereby leading to an additional current component, we talk of the *diffusion current*.

We now consider these two current components separately.

Generation within Space-Charge Region: Generation Current.

In a reverse-biased depletion region, for reverse bias $V_R \gg kT/q$, the concentrations of carriers are reduced well below their equilibrium concentrations. This is because the electric field, which is increased above its equilibrium value by the applied reverse voltage, sweeps the carriers out of the depletion region—it sweeps holes to the *p*-region and electrons to the *n*-region.

Because of the reduction in carrier concentrations, of the four recombination-generation processes that take place through intermediate-level centers (see Figure 5.11), only the *emission processes* are important. The capture processes are not important because their rates are proportional to the concentrations of the free carriers which, as we said above, are very small in a reverse-biased depletion region.

The only way the two emission processes can operate in steady state is by *alternating*; thus the centers within the depletion region alternately emit electrons and holes. The rate of generation of electron-hole pairs in such a situation can be readily obtained from Equation 5.40 by setting $p, n \ll n_i$. This leads to

$$U = -\frac{\sigma_p \sigma_n v_{th} N_t n_i}{\sigma_n e^{(E_t - E_i)/kT} + \sigma_p e^{(E_i - E_t)/kT}} \equiv -\frac{n_i}{2\tau_o} \qquad (6.40)$$

where τ_o is defined as the *effective lifetime within a reverse-biased depletion region* and is given by

$$\tau_o \equiv \frac{\sigma_n e^{(E_t - E_i)/kT} + \sigma_p e^{(E_i - E_t)/kT}}{2\sigma_p \sigma_n v_{th} N_t}. \qquad (6.41)$$

N_t = concentration of gen. resp. rec. centers

To visualize the most important features of this expression, let us consider a simpler case in which $\sigma_p = \sigma_n = \sigma$. For this case, Equation 6.40 reduces to

$$U = -\frac{\sigma v_{th} N_t n_i}{2 \cosh\left(\frac{E_i - E_t}{kT}\right)}. \qquad (6.42)$$

Note that, in agreement with our argument in Chapter 5, *only those centers whose energy level E_t is near the intrinsic Fermi level E_i contribute significantly to the generation rate.* The generation rate falls exponentially as the center level moves away from the middle of the gap in either direction. In the particular case when $E_t = E_i$, τ_o will be the same as the lifetime τ of the carriers in a neutral material, $1/\sigma v_{th} N_t$.

One electron-hole pair generated provides one electronic charge to the external circuit. Thus the magnitude of the current due to generation within the depletion region will be given by

$$I_{\text{gen}} = q |U| W A_J, \qquad (6.43)$$

or

$$I_{\text{gen}} = \tfrac{1}{2} q \frac{n_i}{\tau_o} W A_J \qquad (6.44)$$

where A_J is the cross-sectional area of the *p-n* junction.

Current-Voltage Characteristics

If the centers are indeed located near the intrinsic Fermi level, τ_o will be practically independent of temperature. We would then expect the generation-current component to have the same temperature dependence as n_i. The generation-current component is dependent on the magnitude of the applied reverse bias—at higher biases W is larger, more centers are included within the depletion region, and the generation current increases in proportion to W.

Generation outside the Space-Charge Region: Diffusion Current. In the neutral regions outside the space-charge region there is no significant electric field present and the minority carriers move only by diffusion. If electron-hole pairs are generated in the n-region in the vicinity of the space-charge region, those holes that reach the edge of the space-charge region by diffusion will be swept toward the p-region by the increased electric field present within the space-charge region. Conversely, those electrons that reach the space-charge region edge from the neutral p-region will be swept by the electric field toward the n-region. These current components are referred to as the *diffusion current*. To calculate the magnitude of the diffusion current we have to solve the steady-state diffusion equation for minority carriers. For electrons in the p-region, this is

$$D_n \frac{d^2 n_p}{dx^2} + G_L - \frac{n_p - n_{po}}{\tau_n} = 0. \qquad (6.45)$$

In this equation, in addition to the net thermal generation-recombination term $U = (n_p - n_{po})/\tau_n$, we have also included a term G_L due to external means of generating carriers such as, for instance, by illumination. This equation is the same as the one we used in Chapter 5 in the case dealing with surface recombination. As in that problem, far away from the space-charge region the carrier concentration will be given by

$$n_p(\infty) = n_{po} + \tau_n G_L. \qquad (6.46)$$

At the depletion-region edge, for large enough reverse bias $V_R \gg kT/q$ the minority carrier concentration will be zero, since any minority carriers reaching the depletion region will be rapidly swept away by the field as we have argued above. Thus,

$$n_p(0) = 0 \qquad (6.47)$$

if, for this discussion, we take our origin, $x = 0$, at the edge of the depletion region.

The solution of this boundary value problem is given by

$$n_p(x) = (n_{po} + \tau_n G_L)(1 - e^{-x/L_n}) \qquad (6.48)$$

where $L_n \equiv \sqrt{D_n \tau_n}$ is the diffusion length of electrons in the p-region. The diffusion current due to electrons is then given by

$$I_{\text{diff},n} = (-q)\left(-D_n \frac{dn_p}{dx}\bigg|_{x=0}\right) A_J = qD_n \frac{(n_{po} + \tau_n G_L)}{L_n} A_J. \quad (6.49)$$

A similar argument for the n-region gives the diffusion current due to holes in the n-region reaching the depletion region edge,

$$I_{\text{diff},p} = qD_p \frac{(p_{no} + \tau_p G_L)}{L_p} A_J. \quad (6.50)$$

In the case of *no illumination*, these reduce to

$$I_{\text{diff},n} = qD_n \frac{n_{po}}{L_n} A_J = qD_n \frac{n_i^2}{N_A L_n} A_J \quad (6.51)$$

and

$$I_{\text{diff},p} = qD_p \frac{p_{no}}{L_p} A_J = qD_p \frac{n_i^2}{N_D L_p} A_J, \quad (6.52)$$

respectively, where we have incorporated the equilibrium condition. Note the absence of bias dependence in these expressions—as we have mentioned, they represent the case only for $V_R \gg kT/q$, for which the diffusion current *saturates*. The temperature dependence of the diffusion current is the same as that of n_i^2.

The diffusion-current components can be interpreted also in the following way. Those minority carriers that are generated within a diffusion length of the edge of the depletion region will contribute to the diffusion current because they have a chance to reach the edge of the depletion region. Thus we would expect the diffusion current to be given by

$$I_{\text{diff}} = q \text{ [net generation rate per unit volume in the neutral region]} \times \text{[diffusion length]}\, A_J.$$

To estimate the net generation rate in the neutral region without illumination, we use Equation 5.42 subject to the condition that $p_n \ll p_{no}$ or $n_p \ll n_{po}$. (Note that this is quite different from the condition we applied in the reverse-biased depletion region where both carriers were depleted below n_i. Here we need to say only that the minority carrier concentration near the edge of the depletion region is lower than it would be in equilibrium.) This results in

$$I_{\text{diff},n} = q \left[\frac{n_{po}}{\tau_n}\right] L_n A_J$$

for the example of electrons in the *p*-region. This result is the same as Equation 6.51.

The generation rate per unit volume in the *neutral region* depleted of minority carriers only, p_{no}/τ_p or n_{po}/τ_n, can be drastically different from the generation rate per unit volume within the *depletion region*, $n_i/2\tau_o$. This can lead to very wide differences in the relative importance of the diffusion-current component and the generation-current component of the reverse current. The ratio of the two currents, for $\tau_n = \tau_o = \tau$ is given by

$$\frac{I_{\text{diff},n}}{I_{\text{gen}}} = \frac{\dfrac{n_{po}}{\tau}L_n}{\dfrac{1}{2}\dfrac{n_i}{\tau}W} = 2\frac{n_{po}}{n_i}\frac{L_n}{W} = 2\frac{n_i}{N_A}\frac{L_n}{W}. \tag{6.53}$$

This ratio is evidently proportional to the intrinsic carrier concentration n_i. As the temperature is increased, the diffusion current has an increasing tendency to dominate. Between different materials, those with smaller band gap, hence larger n_i, have a larger diffusion-current-to-generation-current ratio than those with a larger band gap.

These two tendencies are illustrated by the experimental data shown in Figure 6.19. Here the reverse current-voltage characteristics of germanium, silicon, and gallium arsenide diodes are shown at various temperatures. It is evident that the germanium and the silicon diodes exhibit both the diffusion-current component—which is distinguished by its lack of voltage dependence—and the generation-current component—which is distinguished, in contrast, by its voltage dependence—in some temperature range. The only difference is the temperature at which transition from one type of characteristic to the other takes place. The gallium arsenide diode does not reach this transition, but it does show a tendency toward it.

In Figure 6.20, the reverse current at 1 volt reverse bias is plotted as a function of the reciprocal of the absolute temperature for all three diodes. Different types of points are employed here to designate primarily generation-current type characteristics and primarily diffusion-current type characteristics. A break is evident in each set of data, occurring at a higher temperature for the higher band-gap materials, in agreement with our argument. The temperature dependences of the generation and diffusion current components evidently agree reasonably well with the temperature dependence of n_i and of n_i^2, respectively, as expected from Equations 6.44 and 6.51, or 6.52.

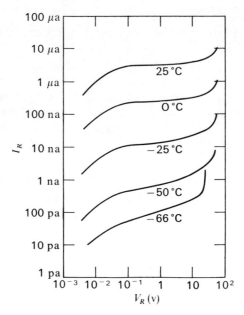

Fig. 6.19a Reverse current-voltage characteristics at various temperatures. Germanium diode.

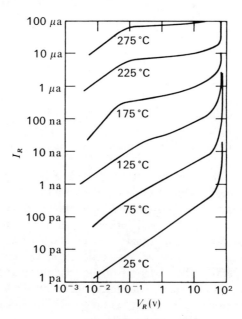

Fig. 6.19b Silicon diode.

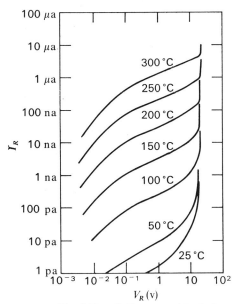

Fig. 6.19c Gallium arsenide diode.

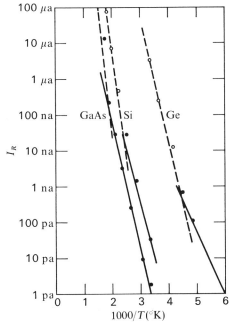

Fig. 6.20 Reverse current at $V_R = 1$ v as a function of temperature. Dots designate primarily generation-current type, circles designate primarily diffusion-current type current-voltage characteristics. Dashed lines represent temperature dependence of n_i^2; solid lines that of n_i.

Let us now consider the reverse current flowing in a diode under illumination. If the illumination is strong, i.e., $G_L \gg n_{po}/\tau_n, p_{no}/\tau_p, n_i/\tau_o$, we can replace the thermal generation rate by G_L in both the generation-current and the diffusion-current components. This leads to

$$I_{\text{photo}} = qG_L(L_p + L_n + W)A_J. \qquad (6.54)$$

Because the sum of the two diffusion lengths is usually larger than the depletion region width, the photocurrent will show little bias dependence.

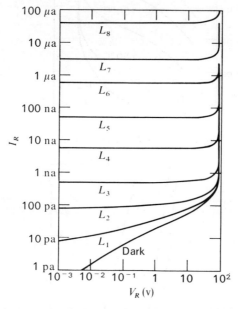

Fig. 6.21 Reverse current of a silicon diode in the dark, and under illumination by white light of various intensities.

This is illustrated in Figure 6.21 where the reverse current of the silicon diode is shown in the dark and under illumination by white light of various intensities.

b. Forward Bias

If a negative voltage is applied to the n-region relative to the p-region, as shown in Figure 6.22, the total variation of the electrostatic potential across the junction, ϕ_T, is reduced. In equilibrium, the condition of zero current flow was established by a precise balance between the diffusion

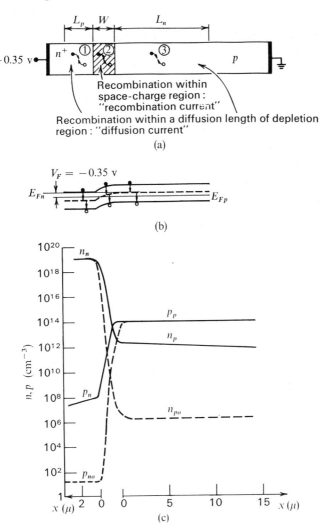

Fig. 6.22 Illustration of the mechanisms giving rise to forward current.
(a) n^+p diode under 0.35-v forward bias.
(b) Corresponding band diagram.
(c) Corresponding distribution of electrons and holes.
Dashed lines indicate equilibrium electron and hole distributions.

and the drift currents of each carrier across the *p-n* junction. Since under forward-bias conditions the total electrostatic potential variation, and with it the electric field across the space-charge region, is reduced, the drift current component of each carrier is also reduced and the balance between drift and diffusion contributions to the current is disturbed, resulting in a net current flow.

It is important to realize that although the potential variation across the junction is reduced, it will not vanish. As a consequence, the electric field in the space-charge region will still be directed toward the *p*-region. Thus the current flow will be *against the direction of the electric field*. This is possible because of the huge electron and hole concentration gradients existing near the junction.

An alternative way of considering the forward-bias case is as follows. With a negative potential applied to the *n*-region relative to the *p*-region, excess electrons will be pushed into the *n*-region and excess holes into the *p*-region. Thus the electron and hole concentrations will both be above their respective equilibrium values and therefore the "*pn*" product will exceed n_i^2 throughout portions of the semiconductor. Under such *injection* conditions the carrier concentrations will attempt to return to their equilibrium values by recombination. In the steady state, electrons and holes disappearing through recombination will be replenished by more electrons and holes coming in through the contacts to the *n* and the *p* regions, respectively. This gives rise to a forward current.

According to this second picture, the *forward current is due to the recombination of electron-hole pairs* in much the same way as the reverse current, which was considered in the previous section, is due to the generation of electron-hole pairs in various regions of the semiconductor. Thus it should be possible to derive the magnitude of the forward current by considering the rates of recombination in the various regions of the semiconductor.

As in the case of reverse current, we again distinguish between three regions, illustrated in Figure 6.22a: the neutral regions 1 and 3 adjacent to the space-charge region and the space-charge region 2 itself. The magnitude of the total forward current will be given by the flux of electrons entering the *n*-region through the left-hand contact. To calculate this flux of electrons, we will add up the three parts of the flux which are consumed by recombination with holes in the three regions mentioned above. Thus the flux of electrons entering through the contact will be given by

$$\frac{I_F}{q} = \text{(number of electrons recombining with holes in neutral } n\text{-region ① per unit time)}$$

+ (number of electrons recombining with holes in space-charge region ② per unit time)

+ (number of electrons recombining with holes in neutral *p*-region ③ per unit time)

Current-Voltage Characteristics

The current components 1 and 3 are called the *diffusion current* as in the reverse-bias case; the current component 2 is called the *recombination current*.

Recombination outside the Space-Charge Region: Diffusion Current. Let us first consider the flux of electrons F_n injected into the neutral *p*-region—i.e., current component 3. Since electrons in the *p*-region are minority carriers, in low-level injection they will move away from the edge of the space-charge region by diffusion only (see Chapter 5). In order to calculate their flux we first must obtain their distribution in the *p*-region.

This distribution in the steady state will be obtained by solving the diffusion equation under steady-state conditions (Chapter 5),

$$D_n \frac{d^2 n_p}{dx^2} - \frac{n_p - n_{po}}{\tau_n} = 0. \tag{6.55}$$

The first boundary condition requires that the concentration of excess electrons far away from the edge of the depletion region should vanish, i.e.,

$$n_p(\infty) = n_{po}.† \tag{6.56}$$

The second boundary condition is that the concentration of injected electrons at the edge of the space-charge region, where we choose our origin, $x = 0$, should equal some constant value $n_p(0)$.

This problem is identical to the case of injection from one boundary considered in Chapter 5. Thus the solution is also identical to the solution in the earlier case and is given by

$$n_p(x) = n_{po} + [n_p(0) - n_{po}] e^{-x/L_n}. \tag{6.57}$$

The flux of electrons entering the *p*-region is then given by

$$F_n = -D_n \frac{dn_p}{dx}\bigg|_{x=0} = D_n \frac{n_p(0) - n_{po}}{L_n}. \tag{6.58}$$

Next, we consider the flux of electrons consumed by recombination with holes in region 1. For every electron recombining in this region, a hole must be injected into the *n*-region from the space-charge region. Thus we can evaluate this contribution to the current by calculating the flux of holes F_p entering the *n*-region. This can be done in a manner entirely analogous to the above calculation of the flux of electrons entering the *p*-region and results in

$$F_p = D_p \frac{p_n(0) - p_{no}}{L_p} \tag{6.59}$$

† This condition assumes that the contact to the *p*-region is "infinitely" far from the junction.

where $p_n(0)$ is the hole concentration at the edge of the space-charge region, on the n-side. In each case, x designates the distance from the edge of the space-charge region, as shown in Figure 6.22.

In order to evaluate these fluxes fully, we need to determine $n_p(0)$ and $p_n(0)$. For this, we must resort to an important simplifying assumption: we assume that *quasi-equilibrium* holds, i.e., that the *"pn" product is a constant* throughout the forward-biased space-charge region. This constant, however, will be larger than n_i^2, the value of the "pn" product in equilibrium.

An alternative way of stating the assumption of quasi-equilibrium is in terms of *quasi-Fermi levels*. The quasi-Fermi level for either carrier is defined as *that quantity which, when substituted into the place of the Fermi level, gives the concentration of that carrier under non-equilibrium conditions.* Thus the concentration of electrons under non-equilibrium conditions is given by

$$n = n_i e^{(E_{Fn}-E_i)/kT} \qquad (6.60)$$

where E_{Fn} is the quasi-Fermi level for electrons; and, likewise, the concentration of holes under non-equilibrium conditions is given by

$$p = n_i e^{(E_i-E_{Fp})/kT} \qquad (6.61)$$

where E_{Fp} is the quasi-Fermi level for holes.

As a consequence of this definition, the fluxes of electrons and of holes now become

$$F_n = -\frac{1}{q} \mu_n n \frac{dE_{Fn}}{dx} \qquad (6.62)$$

and

$$F_p = \frac{1}{q} \mu_p p \frac{dE_{Fp}}{dx}. \qquad (6.63)$$

In the neutral n-region, electrons are transported principally by drift. Thus the flux of electrons in the n-region is given simply by

$$F_n = -\frac{1}{q} \mu_n n \mathscr{E} = -\frac{1}{q} \mu_n n \frac{dE_i}{dx}. \qquad (6.64)$$

Comparison of Equation 6.64 with 6.62 shows that $dE_{Fn}/dx = dE_i/dx$, i.e., that the quasi-Fermi level follows the energy band variation in a parallel fashion. A similar argument leads to an identical conclusion regarding the quasi-Fermi level of holes in the neutral p-region. Thus, *in a neutral region, the quasi-Fermi level for the majority carriers will behave exactly as the Fermi level does in equilibrium.*

Current-Voltage Characteristics

In accordance with this result, we show the quasi-Fermi level for electrons E_{Fn} and that for holes E_{Fp} in the place of the Fermi level in the respective neutral regions of Figures 6.18 and 6.22. In both figures, the band diagram, and consequently the quasi-Fermi level for the majority carriers is shown horizontal within the neutral regions because both bias situations correspond to relatively small current densities.

Because the electrostatic potential variation across the junction is altered in the amount of the applied voltage V_J, it follows that the quasi-Fermi level for electrons in the neutral n-region will also be displaced from the quasi-Fermi level for holes in the neutral p-region by the applied voltage, or

$$E_{Fn} - E_{Fp} = -qV_J \qquad (6.65)$$

as indicated in Figures 6.18 and 6.22.

We can now employ this exceedingly important result in conjunction with the assumption of quasi-equilibrium, i.e., pn = constant throughout the space-charge region, to calculate the concentration of minority carriers at the edges of the space-charge region under forward-bias condition, $n_p(0)$ and $p_n(0)$. It follows from Equations 6.60 and 6.61 that, in quasi-equilibrium, $(E_{Fn} - E_{Fp}) = -qV_J$ throughout the space-charge region also, i.e., both quasi-Fermi levels are extended horizontally across the space-charge region. This is indicated in Figure 6.22.

Thus *the assumption of quasi-equilibrium is equivalent to assuming that the quasi-Fermi levels are constant across the space-charge region.*

The minority carrier concentrations at the respective edges of the space-charge regions are consequently given by

$$n_p(0) = n_i e^{(E_{Fn}-E_i)/kT} = n_{po} e^{q|V_F|/kT} \qquad (6.66)$$

and

$$p_n(0) = n_i e^{(E_i-E_{Fp})/kT} = p_{no} e^{q|V_F|/kT}. \qquad (6.67)$$

Thus the *minority carrier concentrations at the respective edges of the space-charge region will be enhanced by the exponential factor* $e^{q|V_F|/kT}$. The carrier concentrations in a forward-biased one-sided n^+p junction, calculated on the basis of this assumption, are shown in Figure 6.22c.

Noting now that $n_{po} = n_i^2/N_A$ and $p_{no} = n_i^2/N_D$, the equations for the respective diffusion current components become

$$I_{\text{diff},n} = -qD_n \frac{n_i^2}{N_A L_n} [e^{q|V_F|/kT} - 1] A_J \qquad (6.68)$$

and

$$I_{\text{diff},p} = -qD_p \frac{n_i^2}{N_D L_p} [e^{q|V_F|/kT} - 1] A_J. \qquad (6.69)$$

Comparison of these equations with equations 6.51 and 6.52 shows that the diffusion current component under forward bias $|V_F| \gg kT/q$ equals the diffusion current component under reverse bias, multiplied by the exponential factor $e^{q|V_F|/kT}$.

Recombination within the Space-Charge Region: Recombination Current. The electrons lost by recombination in the space-charge region 2 give rise to a current component

$$I_{\text{rec}} = -qA_J \int_0^W U\,dx. \qquad (6.70)$$

This integral is complicated because U depends on n and p, and n and p in turn depend on the distance x in a complicated manner. However, the important features of the result can be illustrated in an approximate fashion.

We saw in Chapter 5 that the most effective recombination centers are those which are located near the middle of the energy gap. Thus we now consider centers with energy level $E_t = E_i$. Furthermore, for simplicity, we take $\sigma_p = \sigma_n = \sigma$. In such a case the recombination rate U is given by

$$U = \sigma v_{th} N_t \frac{pn - n_i^2}{n + p + 2n_i}. \qquad (6.71)$$

Because of the assumption of quasi-equilibrium, the product of electron and hole concentrations throughout the space-charge region is given by

$$pn = n_i^2 e^{q|V_F|/kT}. \qquad (6.72)$$

Thus the recombination rate is

$$U = \sigma v_{th} N_t \frac{n_i^2[e^{q|V_F|/kT} - 1]}{n + p + 2n_i}. \qquad (6.73)$$

For a given forward bias V_F, U will have its maximum value at that location in the space-charge region where the sum of the electron and hole concentrations $(p + n)$ will be at its minimum value. Since the product of these concentrations $pn = $ constant, the condition $d(p + n) = 0$ leads to

$$dp = -dn = \frac{pn}{p^2}dp \qquad (6.74)$$

or

$$p = n \qquad (6.75)$$

as the condition for this minimum. This condition exists at that location within the space-charge region where the intrinsic Fermi level E_i is halfway between the quasi-Fermi level for electrons E_{Fn} and for holes E_{Fp}.

Current-Voltage Characteristics

Here, the carrier concentrations will be given by

$$n = p = n_i e^{q|V_F|/2kT} \tag{6.76}$$

and, therefore, U_{max} by

$$U_{max} = \sigma v_{th} N_t \frac{n_i^2 [e^{q|V_F|/kT} - 1]}{2n_i [e^{q|V_F|/2kT} + 1]}. \tag{6.77}$$

For $|V_F| \gg kT/q$,

$$U_{max} \doteq \tfrac{1}{2} \sigma v_{th} N_t n_i e^{q|V_F|/2kT} \tag{6.78}$$

or

$$U_{max} \doteq \frac{1}{2} \frac{n_i}{\tau_0} e^{q|V_F|/2kT}. \tag{6.79}$$

Note that the maximum recombination rate within the space-charge region under forward bias equals the generation rate under reverse bias multiplied by the exponential factor $e^{q|V_F|/2kT}$.

We can now approximate the recombination current component I_{rec} by

$$I_{rec} \cong -\tfrac{1}{2} q \frac{n_i}{\tau_0} W e^{q|V_F|/2kT} A_J \tag{6.80}$$

which amounts to taking $U = U_{max}$ throughout the space-charge region.†

As in the reverse case, it is interesting to compare the magnitudes of the diffusion and the recombination current components. Using the electron diffusion current component, this ratio is given by

$$\frac{I_{diff,n}}{I_{rec}} = 2 \frac{n_i}{N_A} \frac{L_n}{W} e^{q|V_F|/2kT}. \tag{6.81}$$

As in the reverse-bias case, this ratio depends on the temperature and the band gap of the material through n_i. In addition, it depends on the forward voltage through the exponential factor. For silicon at room temperature, at small forward voltages the recombination current generally dominates and at forward voltages larger than about 0.4 to 0.5 volt the diffusion current usually dominates.

Both current components can also be observed in other materials, but the transition from one characteristic to the other depends on the temperature and the band gap of the material. This is illustrated in Figure 6.23 where we show the forward current of germanium, silicon, and gallium arsenide diodes as a function of forward voltage. We can distinguish the

† Since U decreases exponentially with distance away from the point where it has its maximum value, a better approximation would be to multiply U_{max} by the volume of the space-charge region within which the potential changes by an amount kT/q; that is, by $(kT/q\mathscr{E}) A_J$.

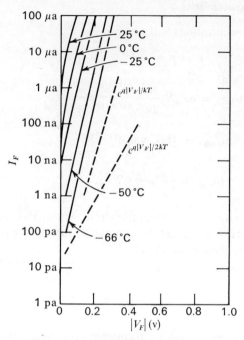

Fig. 6.23a Forward current-voltage characteristics at various temperatures. Germanium diode.

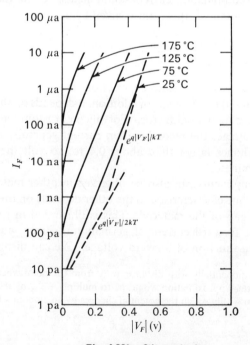

Fig. 6.23b Silicon diode.

188

Current-Voltage Characteristics

recombination current component from the diffusion current component by their different voltage dependences which correspond to slopes of $q/2kT$ and q/kT, respectively, in such a semi-logarithmic representation. These slopes, for room temperature, are shown by the dashed lines. We can observe the change in the slope as the diffusion current begins to dominate with increasing forward bias in the case of the silicon diode.

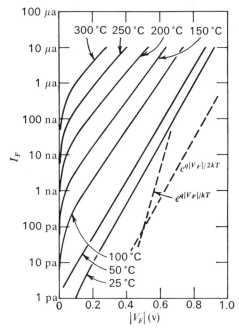

Fig. 6.23c Gallium arsenide diode. Dashed lines indicate slopes of $e^{q|V_F|/kT}$ and $e^{q|V_F|/2kT}$ dependences at 25°C.

This comparison is particularly clear in Figure 6.24 which shows the three room-temperature characteristics together and over a larger current range.

A simple empirical representation of the forward current-voltage characteristics is given by the formula

$$I_F \propto e^{q|V_F|/mkT} \tag{6.82}$$

where the empirical factor $m = 1$ for pure diffusion current and $m = 2$ for pure recombination current, provided the above simple theory is applicable. When both currents are comparable, m will vary between 1 and 2.

A more gradual increase in current at high forward voltages is evident in all three cases shown in Figure 6.24. This phenomenon is associated

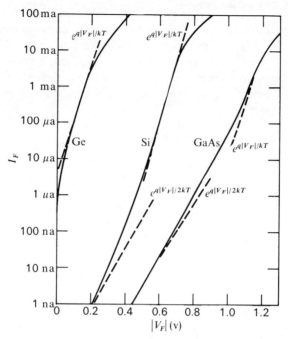

Fig. 6.24 Comparison of the forward current-voltage characteristics of the germanium, silicon, and gallium arsenide diodes at 25°C. Dashed lines indicate slopes of $e^{q|V_F|/kT}$ and $e^{q|V_F|/2kT}$ dependences.

with the onset of high-level injection in which the theory presented in this chapter must be modified. Because high-level injection phenomena are particularly important in the operation of junction transistors, we will discuss them in somewhat greater detail in the next chapter.

c. Diodes of Finite Length (Narrow Base Diodes)

In deriving the distribution of injected minority carriers we have assumed that the contact to the p-region is "infinitely" far from the edge of the depletion region. In many important cases this assumption is not justified. The proximity of the contact to the p-region will change the boundary condition and thereby change the distribution of the injected minority carriers.

The distribution of injected carriers in a case when the separation W_B between a contact and the injecting junction is smaller than the diffusion length L_n (for the case of a p-type substrate) is of paramount importance in transistor operation, and is therefore discussed in detail in the next chapter. For the present we shall merely note that in the limiting case

Junction Breakdown

when $W_B \ll L_n$, the minority carrier distribution will approach the straight-line distribution,

$$n_p(x) = n_{po} + [n_p(0) - n_{po}]\left(1 - \frac{x}{W_B}\right) \qquad (6.83)$$

in contrast to the exponential distribution, Equation 6.57, which holds when $W_B \gg L_n$. As a result, the diffusion length L_n or L_p is replaced by W_B in our formulas for the forward bias diffusion current components, Equations 6.68 and 6.69, as can be readily verified.

6.7 JUNCTION BREAKDOWN[4]

As we discussed earlier, the principal characteristic of a *p-n* junction diode is that it *rectifies*: it permits significant conduction in one direction only. Accordingly, when a diode also begins to conduct in the other direction, its rectifying properties are lost. It is well known that all diodes will conduct in the reverse direction if a sufficiently large reverse voltage is applied to them. This phenomenon is called *junction breakdown*. We shall now discuss the two mechanisms of junction breakdown.

a. Zener Breakdown (Tunneling)

If a high field exists within a semiconductor, for instance, within a reverse-biased depletion region, the covalent bonds between neighboring silicon atoms will be distorted as illustrated schematically in Figure 6.25. If the field becomes high enough, some bonds will be "torn" apart, resulting in conduction electrons and holes. In the band diagram representation, this corresponds to a transition of a valence electron from the valence band to the conduction band. This process, in which an electron penetrates through the energy gap, is called *tunneling*.

Tunneling can take place only if the electric field is very high. Experimentally it is found[5] that the critical field $\mathscr{E}_{\text{crit}}$ at which tunneling becomes probable, i.e., at which *Zener breakdown* commences, is approximately 10^6 v/cm or 100 v/μ.

Although Zener breakdown can be observed in certain *p-n* junctions, most often another breakdown mechanism precedes it. This process is called avalanche breakdown.

b. Avalanche Breakdown

The avalanche or impact ionization breakdown process is illustrated schematically in Figure 6.26. Let us consider this figure step by step. Step 1

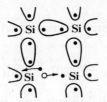

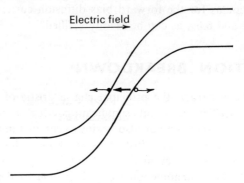

Fig. 6.25 Schematic illustration of the Zener breakdown process.

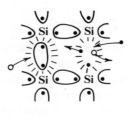

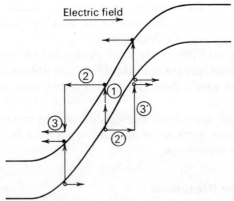

Fig. 6.26 Schematic illustration of the avalanche breakdown process.

Junction Breakdown

is the generation of an electron-hole pair by thermal means, e.g., with the aid of an intermediate level center. If the field within the depletion region is not too high, this process will simply lead to the regular reverse current. However, <u>if the electric field is high enough, the generated electron and hole will gain enough *kinetic energy* before colliding with the lattice so that they will be able to shatter silicon-to-silicon bonds leading to the formation of other electron-hole pairs.</u> In step 2 and 2′, respectively, both the electron and the hole acquire kinetic energy. Step 3 designates the

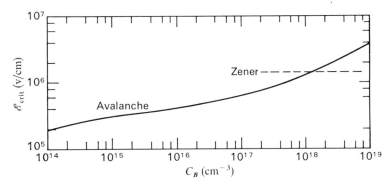

Fig. 6.27 The critical field for avalanche and Zener breakdown in silicon.[5,6]

impact of the electron; the fast-moving electron transfers its kinetic energy to an electron in the valence band, thereby bringing it into the conduction band. The corresponding process, the impact of the energetic hole, is denoted by process 3′. In processes 3 and 3′, *two* new electron-hole pairs were created. All these electrons and holes now begin to accelerate in the high field of the depletion region, as indicated in the figure. In turn, they will again be able to ionize and create other electron-hole pairs in a like manner, and so on. This process is called the *avalanche process*.

If the electric field within the depletion region is not high enough, the carriers cannot acquire sufficient kinetic energy for ionization before suffering a small collision with the lattice. Thus there is a critical field $\mathscr{E}_{crit}$ at which the avalanche process will take place.

Critical field values calculated from measured avalanche breakdown voltages in silicon[6] are shown in Figure 6.27 as a function of the impurity concentration of the substrate of one-sided step junctions. Also shown is the critical field for Zener breakdown.[5] It is evident that Zener breakdown will take place only in materials involving high doping concentrations.

Even before breakdown actually takes place, there will be some *multiplication* of carriers within the depletion region. If without any breakdown-producing mechanism the reverse current is I_{Ro}, then the actual reverse current will be

$$I_R = MI_{Ro} \tag{6.84}$$

where M is the *multiplication factor*. Breakdown of the junction occurs when $M \to \infty$.

It has been found[7] that a good empirical representation of M as a function of reverse voltage V_R for germanium is given by

$$M = \frac{1}{1 - (V_R/BV)^n} \tag{6.85}$$

where n is usually between 3 and 6, its value depending on the semiconductor and the type of the substrate, and BV is the breakdown voltage of the junction. Although this relationship does not hold well in the case of silicon, it can nevertheless be used to illustrate the qualitative features of the multiplication process below avalanche breakdown.

c. Breakdown Voltage of p-n Junctions

We now calculate the breakdown voltage of *p-n* junctions based on the condition that breakdown occurs when the maximum field in the depletion region $\mathscr{E}_{max}$ reaches the critical field $\mathscr{E}_{crit}$. This is only an approximate criterion that can be improved by taking into account the individual ionization rates of electrons and holes, which have been determined experimentally.[4] However, this approximate criterion leads to simple results which are in reasonable agreement with experiment.

For a one-sided step junction, equating the maximum field with the critical field results in

$$\mathscr{E}_{max} = 2\frac{V_R + \phi_B}{W} \doteq 2\frac{V_R}{\sqrt{\dfrac{2K_s\epsilon_0 V_R}{qC_B}}} = \mathscr{E}_{crit} \tag{6.86}$$

where the reverse voltage V_R is now, by definition, the *breakdown voltage* BV of the junction. This leads to the formula for the breakdown voltage of one-sided step junctions:

$$BV = \frac{K_s\epsilon_0 \mathscr{E}_{crit}^2}{2qC_B}. \tag{6.87}$$

For a constant critical field, this shows that the breakdown voltage is inversely proportional to the substrate impurity concentration.

Breakdown voltages of one-sided step junctions in three different materials are shown in Figure 6.28. The curves were calculated[8] by using the experimentally measured individual ionization rates. Also shown are

Junction Breakdown

direct experimental observations[6] of the breakdown voltage of one-sided step junctions in silicon (points). It is evident that in all three cases the breakdown voltage approximately follows the inverse proportionality to substrate impurity concentration predicted by Equation 6.87, especially in the high-voltage range. Furthermore, it is evident that a material with a larger band gap has higher breakdown voltages. This is in agreement with what we would expect on the basis of the simple picture of the

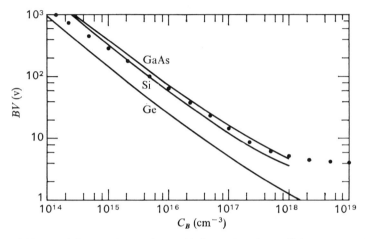

Fig. 6.28 Breakdown voltage of one-sided step-junctions. Points represent direct experimental measurements in silicon;[6] lines were calculated based on ionization rates.[8]

avalanche process discussed above: the carriers must have a higher kinetic energy to be able to ionize across a larger band gap, hence higher breakdown voltages result.

Proceeding in a similar fashion for a linearly graded junction and setting

$$\mathscr{E}_{max} = 1.5 \frac{V_R + \phi_B}{W} \doteq 1.5 \frac{V_R}{\left(\frac{12 K_s \epsilon_0 V_R}{qa}\right)^{1/3}} = \mathscr{E}_{crit} \quad (6.88)$$

leads to the breakdown voltage

$$BV = \sqrt{\frac{32 K_s \epsilon_0 \mathscr{E}_{crit}^3}{9qa}}. \quad (6.89)$$

Thus in the case of linearly graded junctions the important parameter is the concentration gradient at the junction,

$$a \equiv \frac{dC}{dx}\bigg|_{x=x_j}.$$

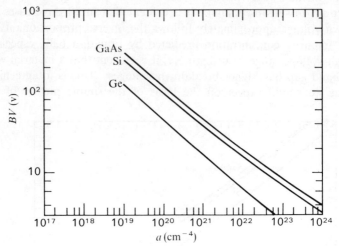

Fig. 6.29 Breakdown voltage of linearly graded junctions.[8]

Curves calculated on the basis of ionization rates are shown in Figure 6.29. It is evident that the breakdown voltage is approximately proportional to $\sqrt{1/a}$ in agreement with Equation 6.89. Thus the simple theory also provides a reasonable description of the breakdown voltage in linearly graded junctions.

Computer calculations[9] of the breakdown voltage of *diffused p-n* junctions, based on ionization rates, are shown in Figure 6.30 as a function of

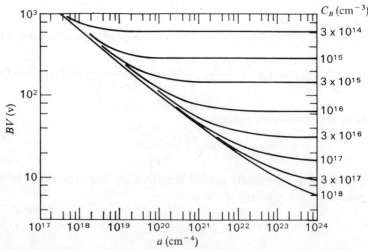

Fig. 6.30 Breakdown voltage of diffused silicon junctions—erfc distribution.[9]

Junction Breakdown

both the substrate impurity concentration C_B and the impurity concentration gradient at the junction a. It is evident that as the gradient a is increased the step-junction limit is approached, whereas for small gradients the linearly graded junction limit holds. The transition point between the two types of behavior depends on the substrate impurity concentration.

In all our discussions of *p-n* junctions so far we have consistently treated

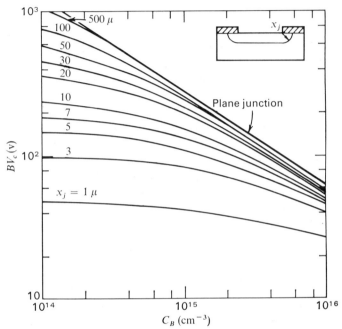

Fig. 6.31 Breakdown voltage of planar silicon one-sided step-junctions showing the effect of junction curvature.[10]

a *p-n* junction as if it was a perfect geometric *plane*. A *planar* junction, i.e., a junction formed by the planar technology, however, is far from being a plane; as we saw in Chapter 3, diffusion from an oxide window proceeds approximately as far along the surface as into the depth of the semiconductor. Because of this, the junction geometry is approximately circular-cylindrical near the window region, as shown in the inset to Figure 6.31. The field and potential distributions in the cylindrical part of the *p-n* junction are different from those in the plane portion of the junction.

In general, the cylindrical part of the junction constitutes a relatively small percentage of the total junction area, and therefore in those characteristics which are cumulative (such as capacitance and current) the error introduced by neglecting these regions is usually relatively small. Breakdown, however, will commence at any region, no matter how small, where the maximum field reaches the critical field of the material. In a *planar p-n* junction the fields will be increased in the cylindrical region because of the finite radius of curvature. Thus the field here will reach the critical field at a lower reverse voltage than that corresponding to the breakdown voltage of the plane region. Hence, this phenomenon will result in a lowering of the breakdown voltage.

Calculations of the breakdown voltage of cylindrical *p-n* junctions,[10] based on the critical-field criterion, are shown in Figure 6.31 with junction depth x_j (which is taken to be the radius of the cylindrical *p-n* junction) as parameter. It is evident that for practical values of the junction depth the breakdown voltage is significantly reduced, especially for low substrate impurity concentrations.

The effect of both substrate concentration and junction depth can be combined into a dimensionless parameter W_c^*/x_j which gives a relative measure of the amount of field crowding that occurs in the cylindrical region of the junction. This parameter is the ratio of the depletion region width within the cylindrical region, at breakdown, W_c^*, to the radius of curvature, x_j. The breakdown theory of cylindrical *p-n* junctions can be rearranged to yield the reduction of breakdown voltage as a function of this dimensionless quantity. This theoretical curve is shown in Figure 6.32

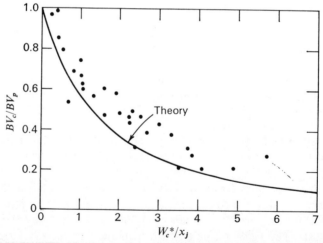

Fig. 6.32 The reduction in breakdown voltage due to the curved shape of *p-n* junctions formed by the planar process. The theoretical curve represents the ratio of breakdown voltages of cylindrical and plane junctions. The experimental points represent the ratio of measured breakdown voltage of planar diodes to the corresponding breakdown voltage of plane diodes.[11]

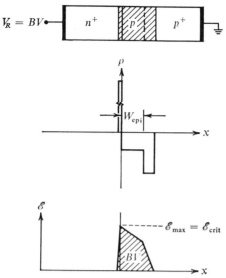

Fig. 6.33 Idealized charge and electric field distribution at breakdown of diodes constructed by epitaxial techniques.

along with numerous experimental data[11] taken on planar *p-n* junctions and compared to their plane counterparts.

Another breakdown voltage limitation may be encountered in diodes constructed by epitaxial techniques such as the one illustrated in Figure 6.1a. This type of construction is often employed in order to minimize the series resistance due to the substrate. In a structure of this kind, if the epitaxial film is relatively thin, the reverse-biased depletion region may reach the highly doped substrate. This is called the *reach-through* condition, illustrated in an idealized manner in Figure 6.33.

An approximate calculation of the area under the idealized electric field distribution, when the maximum field in the depletion region reaches the critical field, leads to the reach-through limited breakdown voltage:

$$BV = \mathscr{E}_{\rm crit} W_{\rm epi} - \frac{qN_A W_{\rm epi}^2}{2K_s \epsilon_0} \qquad (6.90)$$

where $W_{\rm epi}$ is the width of the lowly doped region. Because of the redistribution of the substrate impurity during epitaxial growth and subsequent high temperature heat treatments, $W_{\rm epi}$ will be smaller than the distance to the metallurgical interface (see Chapter 3).

Calculations of the reach-through limited breakdown voltage based on ionization rates[12] are shown in Figure 6.34. Note that for the limit of large

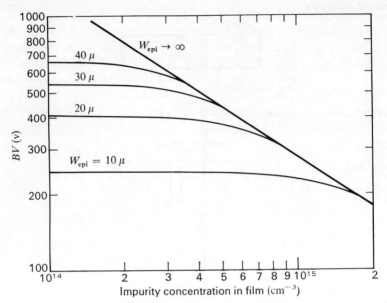

Fig. 6.34 Reach-through limited breakdown voltage of diodes constructed by epitaxial techniques.[12]

epitaxial thickness the regular junction breakdown is reached, but that as the epitaxial film thickness is decreased, the breakdown voltage is also decreased.

We conclude our discussion of breakdown in *p-n* junctions by considering a very important anomalous phenomenon called *soft breakdown*, for want of a better name. A "soft" reverse current-voltage characteristic is illustrated in Figure 6.35 in comparison with the corresponding "hard" breakdown characteristic. It is evident that a large excess reverse current

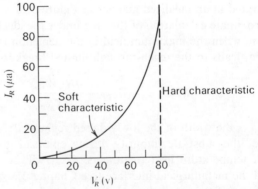

Fig. 6.35 Illustration of the reverse current-voltage characteristics of a diode with "soft" breakdown.

already flows in this diode at reverse voltages well below the avalanche breakdown voltage.

The disastrous implications of this phenomenon on the rectifying properties of diodes are self-evident. The mechanisms giving rise to it, however, are not nearly as clear. It has been shown[13] that precipitates of such metals as copper or iron within the silicon crystal can lead to an increased incidence of soft junctions; and it has also been demonstrated that soft junctions can be "hardened" by a treatment similar to the predeposition of phosphorus. (Such a treatment is called *gettering*.) It has been suggested[13] that the excess current is due to localized breakdown in small high-field regions around metallic precipitates present within the crystal. The treatment with phosphorus is supposed to remove the metallic precipitates, thereby restoring the hard reverse current-voltage characteristics.

6.8 TRANSIENT BEHAVIOR

So far in this chapter we have dealt only with the d-c characteristics of *p-n* junctions. When diodes are used in switching applications, they are alternately forward and reverse biased. Their performance in such applications is determined by the speed with which they can assume a new bias condition. Accordingly, we now briefly consider the factors determining the transient behavior of *p-n* junctions. In particular, we estimate the time required to turn a diode off, i.e., to bring it from a forward-biased condition into a reverse-biased condition.

In Figure 6.36a we indicate the distribution of minority carriers in the lowly doped region of a p^+n diode under forward bias conditions ($t \leqslant 0$), and at various times after the bias is reversed. At $t = 0$, when the applied voltage is reversed, the current is also reversed. Initially, the current flowing in the reverse direction is large because of the presence of the excess minority carriers within the *n*-region. The current continues to remove these minority carriers, and therefore the concentration of minority carriers in the *n*-region decreases until it finally reaches the distribution corresponding to the reverse bias condition. This process is illustrated in Figure 6.36b where we show the forward and reverse currents as a function of time.

A rough estimate of the time required for this transient period to take place can be obtained as follows. Under forward-bias conditions, there is a certain charge density per unit area present within the *n*-region, due to the injected minority carriers. This charge density Q_{p_n} is given by

$$Q_{p_n} \cong \tfrac{1}{2} q p_n(0) X = \frac{I_F X^2}{2 D_p A_J} \qquad (6.91)$$

where we have approximated the minority carrier distribution curve by a triangle of base X. In the case of diodes of finite length (also called narrow-base diodes), the actual distribution in fact follows a straight line and X is simply given by the separation between the contact and the injecting junction W_B. When the contact is infinitely far away, i.e., $W_B \gg L_p$, the distribution is exponential and X can be approximated by the diffusion length L_p.

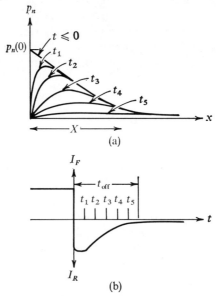

Fig. 6.36 (a) Schematic distribution of minority carriers in the lowly doped region of a p^+n diode at various times after applied bias is reversed.
(b) The corresponding current-time relationship.

If the average reverse current flowing during the turn-off period is $I_{R,\text{ave}}$, the turn-off time will be given by

$$t_{\text{off}} = \frac{Q_{p_n} A_J}{I_{R,\text{ave}}}. \tag{6.92}$$

Combining Equation 6.91 with 6.92 leads to the estimate of the turn-off time,

$$t_{\text{off}} = \left(\frac{I_F}{I_{R,\text{ave}}}\right) \frac{X^2}{2D_p}, \tag{6.93}$$

or

$$t_{\text{off}} = \frac{1}{2}\left(\frac{I_F}{I_{R,\text{ave}}}\right)\tau_p \qquad \text{for } W_B \gg L_p, \tag{6.94}$$

Transient Behavior

and

$$t_{\text{off}} = \frac{1}{2}\left(\frac{I_F}{I_{R,\text{ave}}}\right)\frac{W_B^2}{D_p} \quad \text{for } W_B \ll L_p. \tag{6.95}$$

Thus the turn-off time will depend on the ratio of forward and reverse currents, which are determined by the external circuit,† and on the

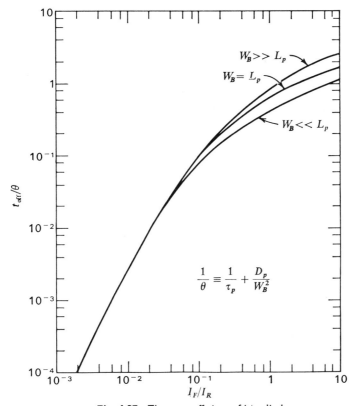

Fig. 6.37 The turn-off time of p^+n diodes.

characteristic time constant of the diode. This characteristic time constant is the lifetime of minority carriers if the contact is infinitely far away, and it is the diffusion time constant W_B^2/D_p if the contact is much closer than a diffusion length.

This simple analysis gives only a rather primitive description of the transient behavior of the diode. The turn-off time can be calculated in a

† The external circuit actually determines the *maximum* value of the reverse current rather than its average.

more exact manner by solving the time-dependent diffusion problem describing the distribution of minority carriers. This has been done in both the general case and its two limits corresponding to wide- and narrow-base diodes.[14] Some results of such calculations are shown in Figure 6.37 which gives the turn-off time as a function of the ratio of the forward current to the maximum reverse current for the wide- and narrow-base limits as well as for the intermediate case of $W_B = L_p$.

REFERENCES CITED

1. H. Lawrence and R. M. Warner, "Diffused Junction Depletion Layer Calculations," *Bell System Tech. J.*, **39**, 389 (1960).
2. J. Hilibrand and R. D. Gold, "Determination of the Impurity Distribution in Junction Diodes from Capacitance-Voltage Measurements," *RCA Rev.*, **21**, 245 (1960).
3. The theory of the current-voltage characteristics of *p-n* junctions was established by W. Shockley, "The Theory of *P-N* Junctions in Semiconductors and *P-N* Junction Transistors," *Bell System Tech. J.*, **28**, 435 (1949). This theory was then extended by C. T. Sah, R. N. Noyce, and W. Shockley, "Carrier Generation and Recombination in *P-N* Junctions and *P-N* Junction Characteristics," *Proc. IRE*, **45**, 1228 (1957). A review of "The Evolution of the Theory for the Voltage-Current Characteristics of *P-N* Junctions" is given by J. L. Moll, *Proc. IRE*, **46**, 1076 (1958).
4. A more detailed discussion of breakdown mechanisms in *p-n* junctions is given in Chapters 11 and 12, J. L. Moll, *Physics of Semiconductors*, McGraw-Hill Book Co., 1964.
5. A. G. Chynoweth, W. L. Feldmann, C. A. Lee, R. A. Logan, G. L. Pearson, and P. Aigrain, "Internal Field Emission at Narrow Silicon and Germanium *P-N* Junctions," *Phys. Rev.*, **118**, 425 (1960).
6. S. L. Miller, "Ionization Rates for Holes and Electrons in Silicon," *Phys. Rev.*, **105**, 1246 (1957).
7. S. L. Miller, "Avalanche Breakdown in Germanium," *Phys. Rev.*, **99**, 1234 (1955).
8. S. M. Sze and G. Gibbons, "Avalanche Breakdown Voltages of Abrupt and Linearly Graded *P-N* Junctions in Ge, Si, GaAs and GaP," *Appl. Phys. Lett.*, **8**, 111 (1966).
9. D. P. Kennedy and R. R. O'Brien, "Avalanche Breakdown Characteristics of a Diffused *P-N* Junction," *IRE Trans. Electron Devices* **ED-9**, 478 (1962).
10. H. L. Armstrong, "A Theory of Voltage Breakdown of Cylindrical *P-N* Junctions, with Applications," *IRE Trans. Electron Devices* **ED-4**, 15 (1957).
11. O. Leistiko and A. S. Grove, "Breakdown Voltage of Planar Silicon Junctions," *Solid-State Electron.*, **9**, 847 (1966).
12. R. J. Whittier, unpublished.
13. A. Goetzberger and W. Shockley, "Metal Precipitates in Silicon *P-N* Junctions," *J. Appl. Phys.*, **31**, 1821 (1960).

Problems

14. R. H. Kingston, "Switching Time in Junction Diodes and Junction Transistors," *Proc. IRE*, **42**, 829 (1954); M. Byczkowski and J. R. Madigan, "Minority Carrier Lifetime in *P-N* Junction Devices," *J. Appl. Phys.*, **28**, 878 (1957); A. S. Grove and C. T. Sah, "Simple Analytical Approximations to the Switching Time in Narrow Base Diodes," *Solid-State Electron.*, **7**, 107 (1964).

PROBLEMS

6.1 (a) Using the potential distribution based on the depletion approximation, derive the electron and hole distributions as a function of distance within the depletion region of a one-sided step junction, in equilibrium.
 (b) Derive expressions for the electron and hole diffusion and drift fluxes crossing the plane where $E_i = E_F$, in equilibrium.
 (c) Calculate the current due to each of the above flux components if the bulk impurity concentration is 10^{16} cm^{-3}, and the junction area is 10^{-3} cm^2. Compare these magnitudes with the forward current data shown in Figure 6.2 and discuss.

6.2 Using the potential distribution based on the depletion approximation, check the validity of the depletion approximation by estimating the width of the region where the carrier concentration is not negligible. Compare this width with the total width of the depletion region under various bias conditions.

6.3 Derive expressions for the electric field, the potential distribution, the built-in voltage, and the depletion region width for a linearly graded junction in equilibrium.

6.4 Derive an expression for the width of the zero-bias depletion region resulting if a contact is formed between a metal and an *n*-type semiconductor. Assume that at the metal-semiconductor interface the conduction band of the semiconductor will be fixed at an energy Φ above the Fermi level of the metal where $\Phi > (E_c - E_F)$ in the bulk of the semiconductor. Compare with the expression for a one-sided step junction. Discuss what happens if $\Phi < (E_c - E_F)$ in the bulk of the semiconductor.

6.5 A diode is made as follows:
 Starting material: 0.175 Ω cm *n*-type silicon.
 Boron predeposition: results in $Q = 10^{15}$ boron atoms/cm^2.
 Drive-in diffusion: 1 hour at 1200°C.
 (a) If the junction area of the diode is 10^{-3} cm^2, calculate its capacitance at zero and at 10 v reverse bias, using the one-sided step junction and the linearly graded junction approximations. (Is the latter meaningful?) Which approximation is best in each case and why?
 (b) Define a voltage range where one or the other approximation can be expected to lead to good results.

6.6 Assuming that both electrons and holes within a reverse-biased depletion region move with a constant maximum drift velocity, $v_{\text{lim}} = 10^7$ cm/sec, construct the electron and hole distribution corresponding to Figure 6.18. Assume that the bulk impurity concentration is 10^{16} cm^{-3} and that the lifetime is 1 μsec. Do this:
 (a) For a silicon diode at room temperature.
 (b) For a germanium diode at room temperature.

6.7 Derive and calculate the small-signal conductance dI_R/dV_R at $V_R = 10$ v for the diodes of the previous problem, at room temperature. Compare with the data shown in Figure 6.19b.

6.8 Using the potential distribution based on the depletion approximation, derive the electron and hole concentration distribution within the space-charge region of a one-sided step-junction under small forward bias. Compare your result with the calculations shown in Figure 6.22.

6.9 Derive an expression giving forward current of a diode as a function of the difference of the applied voltage and the built-in voltage of the junction.

6.10 By equating the expressions giving the electron and hole fluxes in terms of the gradients of the respective quasi-Fermi levels to the formula giving the forward current, arrive at an order-of-magnitude justification of the assumption of quasi equilibrium.

6.11 Express the ratio of the diffusion current component to the recombination current component as a function of forward current instead of forward voltage. Using this expression, compare diodes made of germanium, silicon, and gallium arsenide with respect to the relative importance of the two components. Compare your conclusion with the data shown in Figure 6.24.

6.12 Derive the small-signal conductance dI_F/dV_F of a p-n junction diode, considering both components of the forward current. Calculate its value for the diodes described in Problem 6.6 at $V_J = 0$, and at $V_F = 0.4$ v, at room temperature. Compare with the experimental measurements shown in Figure 6.24. Also, compare with the results of Problem 6.7.

6.13 Derive the forward voltage as a function of temperature at a given current density. Comment on the use of a p-n junction as a thermometer.

6.14 The difference in the breakdown voltage of those planar and plane diffused p-n junctions which are well described by the linearly graded junction approximation is much less than the difference in cases where the junctions follow the one-sided step-junction approximation. Give a simple explanation.

6.15 Based on the simple picture of the avalanche process given in the text, propose a qualitative argument giving the sign of the temperature dependence of the avalanche breakdown voltage.

6.16 Derive an expression giving the time required to turn a diode *on*, *i.e.*, to bring it into forward bias from a reverse-bias condition.

TABLE 6.1
IMPORTANT FORMULAS FOR ONE-SIDED STEP JUNCTIONS

Built-in voltage	$\phi_B \cong 2\dfrac{kT}{q}\ln\dfrac{C_B}{n_i}$
Depletion region width	$W = \sqrt{\dfrac{2K_s\epsilon_0[\phi_B \pm \lvert V_J \rvert]}{qC_B}}$ where $\begin{array}{l}+:\text{reverse}\\-:\text{forward}\end{array}\Big\}$ bias
Maximum electric field	$\mathscr{E}_{\max} = 2\dfrac{\phi_B \pm \lvert V_J \rvert}{W}$
Capacitance per unit area	$C = \dfrac{K_s\epsilon_0}{W}$
Reverse current	$I_R = I_{\text{gen}} + I_{\text{diff}}$ $I_{\text{gen}} = \tfrac{1}{2}q\dfrac{n_i}{\tau}WA_J$ $I_{\text{diff}} = qD\dfrac{n_i^2}{C_B L}A_J$
Forward current	$I_F = I_{\text{rec}} + I_{\text{diff}}$ $I_{\text{rec}} = -\tfrac{1}{2}q\dfrac{n_i}{\tau}W e^{q\lvert V_F\rvert/2kT}A_J$ $I_{\text{diff}} = -qD\dfrac{n_i^2}{C_B L}e^{q\lvert V_F\rvert/kT}A_J$
Avalanche breakdown voltage	$BV = \dfrac{K_s\epsilon_0 \mathscr{E}_{\text{crit}}^2}{2qC_B}$

- **PRINCIPLES OF TRANSISTOR ACTION**
- **CURRENTS FLOWING IN A TRANSISTOR; CURRENT GAIN**
- **LIMITATIONS AND MODIFICATIONS OF THE SIMPLE THEORY**
- **BASE RESISTANCE**
- **MAXIMUM VOLTAGES**
- **MINIMUM VOLTAGES**
- **THERMAL LIMITATIONS**

7

Junction Transistors

The single most important solid-state device is the junction transistor. Its invention brought about an unprecedented growth of research and development work in solid-state physics and engineering. It is the active device on which most discrete and integrated solid-state circuits are based. In this chapter we examine the basic characteristics of junction transistors.

A planar silicon *pnp* transistor is illustrated schematically in Figure 7.1a. The fabrication of such a transistor begins the same way as the fabrication of planar diodes: usually a lightly doped epitaxial film is grown upon a heavily doped substrate of the same type, the surface of the wafer is oxidized, windows are opened in the oxide, and an impurity is permitted to diffuse through these windows to form a junction within the epitaxial film. At this point, however, the surface of the silicon is oxidized again, windows are opened at different places, and another impurity, of the same type as the substrate, diffuses into the silicon to form a second junction. Windows are then opened and metallic contacts are deposited upon both diffused regions. The wafer is then cut apart by scribing, leads are attached to all three regions of the transistor, and the transistor is packaged.

For simplicity, we study an idealized one-dimensional model of this planar transistor analogous to the idealized model of the *p-n* junction diode of the previous chapter. This one-dimensional model, shown in Figure 7.1b, can be considered as a section of the planar transistor along the dashed lines. It neglects the variation of the impurity concentration within each of the three regions of the transistor. Where the concentration variation or the characteristic shape of planar junctions leads to significant deviations, we modify our discussion accordingly.

We begin with a qualitative discussion of the principles of transistor action. We then consider the various current components which flow within a transistor and the factors influencing the current gain by using a

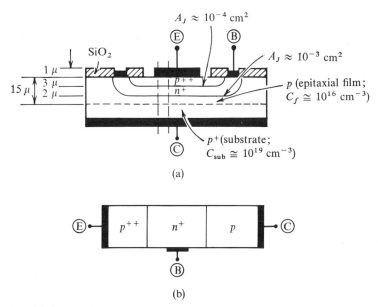

Fig. 7.1 (a) A typical planar *pnp* transistor. Representative impurity concentrations and dimensions are indicated.
(b) Idealized one-dimensional transistor model.
    Ⓔ: Emitter
    Ⓑ: Base
    Ⓒ: Collector

simple theory. The various ways this theory must be modified under certain conditions are discussed next. Then we consider base resistance, the maximum and minimum voltage limitations of transistors, and thermal limitations.

7.1 PRINCIPLES OF TRANSISTOR ACTION

The energy bands pertaining to the idealized *pnp* transistor of Figure 7.1b are shown in Figure 7.2. Figure 7.2a shows the energy bands under equilibrium conditions, i.e., when all three leads labeled Ⓔ, Ⓑ, and Ⓒ, for *emitter*, *base*, and *collector*, are connected together. Under such conditions, as we saw in Chapter 6, the electrostatic potential varies from each semiconductor region to the next in such a way as to balance out the diffusion flux of electrons and holes due to their concentration gradients.

When external voltages are applied to the various regions of the transistor, the junctions may become forward or reverse biased. The

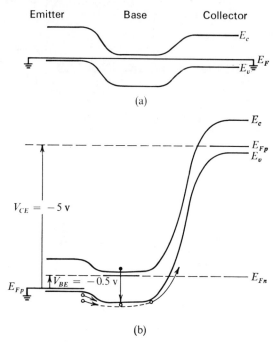

Fig. 7.2 Energy band diagram for a *pnp* transistor.
(a) Equilibrium case.
(b) Under bias.

energy bands under the most important type of bias condition are shown in Figure 7.2b. Under this condition, the junction between emitter and base is forward biased and the junction between base and collector is reverse biased.

Due to the forward biasing of the emitter-base junction, large numbers of holes will now be injected into the *n*-type base. If the two junctions are sufficiently close to each other, most of these holes will reach the collector-base junction where they will be swept across by the electric field. Thus they will be collected in the *p*-type collector. It should be noted that even though the collector-base junction is reverse biased, a large current which approximately equals the forward-biased current of the emitter-base

Principles of Transistor Action

junction will now flow in it. This is the principal feature of transistor action: *a large current flows in a reverse-biased junction due to the existence of a forward-biased junction in its vicinity.*

It should be realized that not all injected holes will reach the collector-base depletion region but that some will recombine with electrons en route through the base. Also, when the emitter-base junction is forward biased, some electrons will be injected into the emitter along with the injection of holes into the base. Finally, electrons and holes will recombine in the space-charge region of the emitter-base junction. Due to these processes, electrons will flow into the base through the base lead. This electron flow, of course, corresponds to a current flowing *out* the base lead as indicated in Figure 7.3.

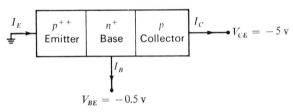

Fig. 7.3 Currents flowing in a *pnp* transistor.

As we have discussed above, the total emitter current I_E consists of those holes which reach the collector, I_C, and of the electrons which flow into the transistor through the base lead, I_B. Thus,

$$I_E = I_C + I_B. \tag{7.1}$$

Two quantities of great importance in the characterization of transistors are the so-called *common-base current gain* α, also referred to as h_{FB}, defined by

$$\alpha \equiv h_{FB} \equiv \frac{I_C}{I_E}, \tag{7.2}$$

and the so-called *common-emitter current gain* β, also referred to as h_{FE}, defined by

$$\beta \equiv h_{FE} \equiv \frac{I_C}{I_B}. \tag{7.3}$$

It is evident from Equation 7.1 that these are related to each other by

$$\beta = \frac{\alpha}{1-\alpha}. \tag{7.4}$$

Under normal operating conditions such as we discussed above, α will

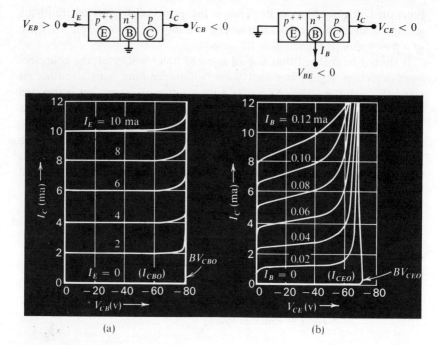

Fig. 7.4 Current-voltage characteristics of a silicon *pnp* transistor. This device is used for illustration throughout this chapter. Its structural parameters are approximately as indicated in Fig. 7.1a.
(a) Common base configuration.
(b) Common emitter configuration.

always be smaller than unity, although its value in a good transistor will approach unity very closely. Accordingly, β will be large in a good transistor.

The current-voltage characteristics of a planar silicon *pnp* transistor whose structural parameters are approximately as indicated in Figure 7.1a are shown in Figure 7.4. This device will be used for illustration throughout this chapter. The characteristics are shown in two biasing configurations. In one, the voltages are applied with respect to the base. This is referred to as the *common-base* mode. In the other configuration, the voltages are applied with respect to the emitter. This is referred to as the *common-emitter* mode.

It is evident from Figure 7.4a that α is in fact very close to unity, and from Figure 7.4b that $\beta \approx 60$ at $I_C = 4$ ma, at low values of V_{CE}. Focusing our attention on the common-emitter configuration, we can also see that a change in I_B brings about a change in I_C which is also about

Principles of Transistor Action

60 times larger. The relevant *small-signal current gain* h_{fe} is defined by

$$h_{fe} \equiv \frac{dI_C}{dI_B}. \tag{7.5}$$

It can be readily shown from the two definitions, Equations 7.3 and 7.5, that the relationship between the small-signal current gain h_{fe} and the d-c current gain h_{FE} is given by

$$h_{fe} = \frac{h_{FE}}{1 - \dfrac{I_C}{h_{FE}} \dfrac{dh_{FE}}{dI_C}}. \tag{7.6}$$

Thus, for a transistor whose current gain h_{FE} is independent of the collector current, $h_{fe} = h_{FE}$.

We now consider the behavior of our transistor in a simple circuit shown in Figure 7.5a. For a given input voltage V_{BE}, a certain d-c base current I_B

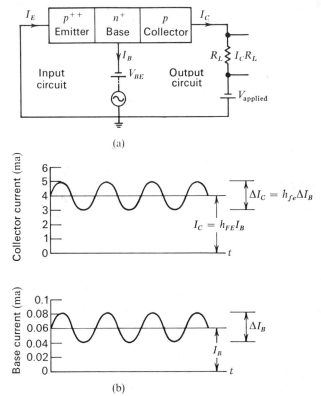

Fig. 7.5 Illustration of the use of the transistor as an amplifier.

and collector current I_C flow in the transistor. If a small a-c signal is now superimposed on the input voltage, the base current will vary as a function of time as illustrated in Figure 7.5b. This variation, in turn, brings about a corresponding a-c variation in the output current I_C which however is h_{fe} times larger than the input current variation. Thus the transistor *amplifies* the input signal.

A transistor can also be used as a switching device. By controlling a small current (the base current), a larger one (the collector current) can be turned on and off.

The first of these applications, as an amplifier, involves small-signal a-c phenomena. The second, as a switch, involves large-signal transient phenomena. However, both of these are consequences of the d-c characteristics and limitations of the junction transistor, to which we devote the remainder of this chapter.

At this point we should define the established terminology of junction transistor parameters. Since the transistor is a three-terminal device, currents and voltages are often specified with three-letter subscripts. The first two of these letters designate the two terminals between which the current or voltage is measured. The third letter designates the state of the third terminal with respect to the second. For example, BV_{CBO} designates the breakdown voltage between collector and base with the emitter-base junction *open*. BV_{CES} designates the breakdown voltage between collector and emitter with base *shorted* to emitter. I_{CBO} and I_{CES} designate the corresponding leakage currents, and so on.

7.2 CURRENTS FLOWING IN A TRANSISTOR; CURRENT GAIN[1]

a. Current Components

The various current components involved in transistor action are illustrated in Figure 7.6 where the current carried by holes I_p is shown as a function of the distance from the emitter through the base to the collector region, for a *pnp* transistor. In the p^+ emitter most of the current is carried by holes so that the hole current I_p equals the total emitter current I_E. As we proceed toward the collector, an increasing fraction of the total current is carried by electrons.

Let us consider the boundary between the emitter-base depletion region and the *n*-type base, designated by the plane $x = 0$. At this boundary, we show the fraction of the emitter current due to holes diffusing into the base $I_{\text{diff},B}$. The rest of the emitter current is carried by electrons. We can separate the electron current at $x = 0$ into two components: component 1 is due to electrons which are injected into the p-type emitter region, while component 2 is due to electrons which are injected into the emitter-base space-charge region where they recombine with holes. As we now proceed toward the collector region, the fraction of the current carried by holes decreases because some of the holes injected to the n-type

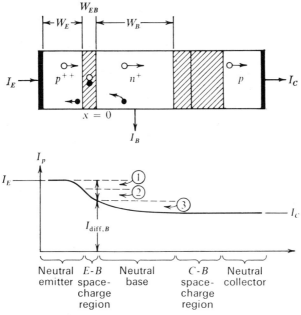

Fig. 7.6 The variation of hole current as a function of position for a pnp transistor.

base recombine with electrons. The fraction of hole current consumed by recombination in the neutral base region is designated as the current component 3. The sum of the three components is the base current.

The fraction of the total current carried by holes will not change any more beyond the base-collector depletion-region boundary. This is because in a reverse-biased depletion region the recombination process is negligible. Similarly, in the neutral collector region the hole current is a majority carrier current and it is not diminished by recombination.

Let us now consider these current components in a quantitative manner. In low-level injection, the transport of the minority carriers injected into the base can be described entirely by diffusion as we saw in Chapter 5. To obtain their distribution, we must solve the steady-state diffusion equation:

$$D_p \frac{d^2 p_n}{dx^2} - \frac{p_n - p_{no}}{\tau_p} = 0 \qquad (7.7)$$

subject to the boundary conditions
$$p_n(0) = p_{no}e^{qV_{EB}/kT} \tag{7.8}$$
and
$$p_n(W_B) = 0, \tag{7.9}$$
where the equilibrium minority carrier concentration within the base is $p_{no} = n_i^2/N_{DB}$, N_{DB} denoting the donor concentration within the base region (assumed to be uniform in our idealized model). The first of these conditions states that the concentration of the minority carriers at the edge of the emitter-base junction space-charge region is increased above their equilibrium value by the exponential factor $e^{qV_{EB}/kT}$, as discussed in Chapter 6. The second boundary condition states that the concentration of minority carriers at the edge of the base-collector junction space-charge region will be zero. This is because this junction is reverse biased; the electric field will immediately sweep across any minority carriers that arrive there.

The solution is

$$p_n(x) = p_{no}\left[1 - \frac{\sinh\frac{x}{L_p}}{\sinh\frac{W_B}{L_p}}\right] + [p_n(0) - p_{no}]\frac{\sinh\frac{W_B - x}{L_p}}{\sinh\frac{W_B}{L_p}}$$

$$\doteq p_n(0)\frac{\sinh\frac{W_B - x}{L_p}}{\sinh\frac{W_B}{L_p}}, \quad \text{for } V_{EB} \gg kT/q. \tag{7.10}$$

Calculations based on this solution are shown in Figure 7.7 where the normalized minority carrier concentration is shown as a function of distance for a fixed diffusion length $L_p = 10\ \mu$ and various values of the base width W_B. It is evident that for $W_B \gg L_p$ the distribution approaches the simple exponential distribution discussed in the previous chapter, Equation 6.57. In the other extreme, where $W_B \ll L_p$, the distribution approaches the simple straight-line form given by

$$p_n(x) = p_n(0)\left(1 - \frac{x}{W_B}\right). \tag{7.11}$$

This limiting straight-line distribution was discussed in Chapter 6 in connection with diodes of finite length.

We can now write down the various components of the total emitter current I_E. By referring to the plane $x = 0$ in Figure 7.6, we see that it will, first of all, consist of the diffusion current of holes injected into the

Currents Flowing in a Transistor; Current Gain

base, $I_{\text{diff},B}$. In good transistors $W_B \ll L_p$, hence we can approximate the distribution of minority carriers within the base by the straight-line distribution, Equation 7.11. This yields

$$I_{\text{diff},B} \doteq qD_{pB} \frac{n_i^2}{N_{DB}W_B} e^{qV_{EB}/kT} A_J \qquad (7.12)$$

where D_{pB} denotes the diffusivity of holes in the base region. The diffusion current of electrons injected into the emitter (component 1 of Figure 7.6) is given by

$$I_{\text{diff},E} = qD_{nE} \frac{n_i^2}{N_{AE}W_E} e^{qV_{EB}/kT} A_J \qquad (7.13)$$

where we have assumed that the emitter depth W_E is much smaller than the diffusion length of electrons in the emitter region. N_{AE} denotes the

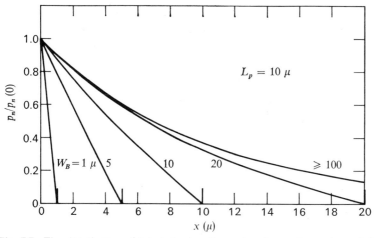

Fig. 7.7 The distribution of injected minority carriers for various values of the base width.

acceptor concentration in the emitter, assumed to be uniform in our idealized model; and D_{nE} is the diffusivity of electrons in the emitter. Finally, the current carried by electrons injected into the emitter-base space-charge region where they recombine with holes (current component 2 in Figure 7.6) is given by

$$I_{\text{rec}} = \tfrac{1}{2}q \frac{n_i}{\tau_o} W_{EB} e^{qV_{EB}/2kT} A_J \qquad (7.14)$$

where W_{EB} is the width of the space-charge region of the emitter-base junction.

An efficient emitter is one in which the components 1 and 2 of Figure 7.6 are small. Thus we will define the *emitter efficiency* by

$$\gamma \equiv \frac{\text{Diffusion current in base}}{\text{Total emitter current}}$$

or

$$\gamma \equiv \frac{I_{\text{diff},B}}{I_E} = \frac{I_{\text{diff},B}}{I_{\text{diff},B} + I_{\text{diff},E} + I_{\text{rec}}}. \tag{7.15}$$

Using Equations 7.12 to 7.14, this becomes

$$\gamma = \frac{1}{1 + \dfrac{N_{DB}W_B}{D_{pB}}\left(\dfrac{D_{nE}}{N_{AE}W_E} + \dfrac{W_{EB}/\tau_0}{2n_i e^{qV_{EB}/2kT}}\right)} \tag{7.16}$$

which can be rearranged to yield

$$\gamma = \frac{1}{1 + \dfrac{B}{E} + \dfrac{1}{2}\sqrt{\dfrac{qBA_J}{I_{\text{diff},B}}}R} \tag{7.17}$$

where

$$\left. \begin{array}{l} B \equiv \dfrac{N_{DB}W_B}{D_{pB}} \\[6pt] E \equiv \dfrac{N_{AE}W_E}{D_{nE}} \\[6pt] R \equiv \dfrac{W_{EB}}{\tau_0} \end{array} \right\} \tag{7.18}$$

The first of these three quantities is the *base factor* which depends on the total number of impurities in the base region. The quantity E, the *emitter factor*, depends on the total number of impurities in the emitter region. Finally, the quantity R, the *recombination factor*, is characteristic of the recombination rate in the emitter-base junction space-charge region.† These and other important formulas relevant to junction transistors are summarized in Table 7.1 at the end of this chapter.

By the definition of the emitter efficiency, the current carried by minority carriers injected into the base is given by γI_E. What is of principal importance is the fraction of this current which reaches the collector-base depletion region and is collected there. The quantity designating the

† This treatment neglects surface recombination. As discussed in Chapter 10, its effect on γ can be approximated by setting

$$R = \frac{W_{EB}}{\tau_0} + s_0 \frac{A_s}{A_J}$$

where $s_0 = \sigma v_{th} N_{st}$ and A_s is the depleted surface area.

Currents Flowing in a Transistor; Current Gain

fraction of the injected current that does reach the collector-base depletion region is called the *transport factor* and is defined by

$$\alpha_T \equiv \frac{\text{Hole current reaching collector}}{\text{Hole current injected into base}}$$

for the case of a *pnp* transistor. In Figure 7.6, this variation of hole current across the base region is indicated by the decrease of hole current marked by 3.

The transport factor can be calculated by using the solution of the minority carrier distribution within the base region, Equation 7.10. By the definition of the transport factor,

$$\alpha_T = \frac{\left.\dfrac{dp_n}{dx}\right|_{x=W_B}}{\left.\dfrac{dp_n}{dx}\right|_{x=0}}. \tag{7.19}$$

Using Equation 7.10, this can be shown to lead to

$$\alpha_T = \frac{1}{\cosh \dfrac{W_B}{L_{pB}}} \tag{7.20}$$

where L_{pB} is the diffusion length of minority carriers in the base region.

For good transistors, W_B is much smaller than L_{pB}. Therefore, to a good approximation,

$$\alpha_T \cong 1 - \frac{1}{2}\left(\frac{W_B}{L_{pB}}\right)^2. \tag{7.21}$$

b. D-C Current Gain

As we have discussed, the two figures-of-merit which are used to describe the performance of transistors in the common-base and common-emitter configurations are the current gains α and β, respectively.

By its definition,

$$\alpha \equiv h_{FB} \equiv \frac{I_C}{I_E} = \gamma \alpha_T.\dagger \tag{7.22}$$

† The relationship $\alpha \equiv I_C/I_E = \gamma \alpha_T$ holds only if the reverse-bias leakage current of the collector-base junction I_{CBO} is negligible. If this is not the case, the relationship becomes

$$\alpha \equiv \frac{I_C}{I_E} = \gamma \alpha_T + \frac{I_{CBO}}{I_E}.$$

Correspondingly,

$$\beta \equiv \frac{I_C}{I_B} = \left(\frac{\gamma \alpha_T}{1 - \gamma \alpha_T}\right)\left(1 + \frac{I_{CBO}}{\gamma \alpha_T I_B}\right).$$

Accordingly, the common emitter current gain will be given by

$$\beta = h_{FE} = \frac{\gamma \alpha_T}{1 - \gamma \alpha_T}. \qquad (7.23)$$

Thus it is clear that we would like the product of the transport factor α_T and the emitter efficiency γ to approach unity as closely as possible.

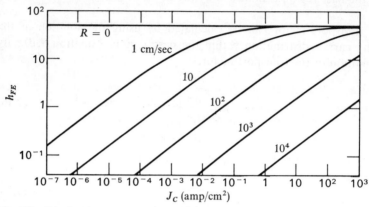

Fig. 7.8 Calculated common-emitter current gain as a function of collector current density for various space-charge region recombination rates, for a silicon transistor. Recombination in neutral base region is neglected, i.e., $\alpha_T = 1$. ($E = 5 \times 10^{15}$ sec/cm^4; $B = 1 \times 10^{14}$ sec/cm^4)

For a transistor with a large β, we can write

$$\frac{1}{\beta} \doteq 1 - \gamma\alpha_T \doteq \underbrace{\frac{1}{2}\left(\frac{W_B}{L_{pB}}\right)^2}_{\substack{\text{Recom-}\\\text{bination}\\\text{within}\\\text{neutral}\\\text{base}}} + \underbrace{\frac{N_{DB}W_B}{D_{pB}} \frac{D_{nE}}{N_{AE}W_E}}_{\substack{\text{Injection}\\\text{into}\\\text{emitter}}} + \underbrace{\frac{N_{DB}W_B}{D_{pB}} \frac{W_{EB}/\tau_o}{2n_i e^{qV_{EB}/2kT}}}_{\substack{\text{Recombination}\\\text{within E-B}\\\text{space-charge}\\\text{region}}}$$

(7.24)

where the origin of each of the three terms is indicated.

In many applications it is important not only that β be large but also that it should not vary with the current level. The nature of the variation of the current gain with collector current is illustrated in Figure 7.8. Here calculations based on Equations 7.17 and 7.23 are shown. These calculations were made by using emitter and base factors approximating those of the *pnp* transistors of Figure 7.1, for various values of the recombination parameter, R. In these calculations we also assumed that the transport

Currents Flowing in a Transistor; Current Gain

factor $\alpha_T = 1$; thus $I_{\text{diff},B} = I_C$. It is evident that in the absence of recombination within the emitter-base space-charge region ($R = 0$), the current gain is independent of the collector current. The larger the recombination rate R, the more the current gain drops at low current levels.

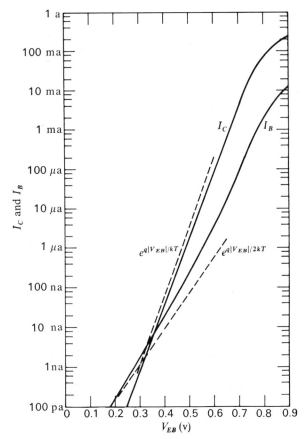

Fig. 7.9 Collector and base currents as a function of emitter-base forward bias for the pnp transistor. ($V_{CB} = 0$)

Experimental measurements of the collector and base currents of the pnp transistor are shown in Figure 7.9 as a function of emitter-base forward bias. Several features should be noted in this figure. First, the collector current follows the voltage dependence predicted for the injected current, Equation 7.12, over eight decades of current. This agreement is an indication of the validity of the injected carrier concentration formulas given in Chapter 6. On the other hand, the base current as a function of emitter-base forward voltage follows

$$I_B \propto e^{qV_{EB}/mkT}$$

with $m \simeq 1.7$ in the lower current range. This is an indication of the fact that most of the base current at low current levels is due to recombination in the emitter-base space-charge region.

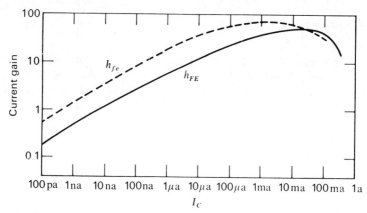

Fig. 7.10 Common-emitter current gain versus collector current for the pnp transistor. ($V_{CB} = 0$)

The common emitter current gain for the same device is shown in Figure 7.10 as a function of collector current. Note that the variation of h_{FE} with collector current is quite similar to the theoretical calculations shown in Figure 7.8, except for the decrease of h_{FE} at high current levels. When the emitter junction area, $\sim 10^{-4}$ cm², is taken into account, a reasonable fit is obtained between theory and experiment for $R \simeq 10$ cm/sec.

Also shown in this figure is the experimentally measured small-signal current gain h_{fe} of this device.

7.3 LIMITATIONS AND MODIFICATIONS OF THE SIMPLE THEORY

a. Transit-Time Limitation

Although it is beyond the scope of this book to discuss the frequency response of transistors, we shall develop a simple estimate of the maximum frequency up to which a transistor can be expected to be useful. This limitation is provided by the time required for the rearrangement of minority carriers in the base region.

Limitations and Modifications of the Simple Theory

If a transistor is to be useful, it is expected that upon a change in emitter-base forward bias the collector current will change also. In order to change the collector current, the minority carrier distribution in the base must be altered as illustrated in Figure 7.11. We will estimate the time required for such a rearrangement by calculating the time needed for holes to travel across the base region. (This calculation is a good example of similar transit-time calculations in other physical systems.)

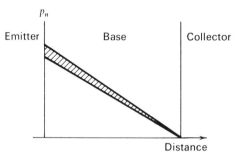

Fig. 7.11 The small-signal variation of minority carrier distribution in the base region.

The distance traveled by a hole in a time dt is given by

$$dx = v(x)\, dt \qquad (7.25)$$

where $v(x)$ is the velocity of a hole. Thus the transit time for holes across the base region of a *pnp* transistor will be given by

$$t_{tr} = \int_0^{W_B} \frac{dx}{v(x)}. \qquad (7.26)$$

The velocity of the holes is related to the hole current and the hole distribution in the base by

$$I_p = qv(x)p(x)A_J. \qquad (7.27)$$

Using a straight-line hole distribution, we can then readily show that the transit time across the base will be

$$t_{tr} = \frac{W_B^2}{2D_{pB}}. \qquad (7.28)$$

The frequency limitation corresponding to this time limit is given approximately by the reciprocal of the transit time.† Experimental measurements of the small-signal common-emitter current gain h_{fe} of the *pnp* transistor as a function of frequency are shown in Figure 7.12. Note that h_{fe} drops

† More rigorous considerations yield $1/2\pi t_{tr}$ for this frequency limitation.

below unity at a frequency of the same order of magnitude as given by the above criterion.

It is interesting to compare the transit time with the base transport factor α_T given earlier. It is evident from Equations 7.21 and 7.28 that the base transport factor is related to the ratio of the transit time across the base to the lifetime by

$$\alpha_T = 1 - \frac{t_{tr}}{\tau_p}. \tag{7.29}$$

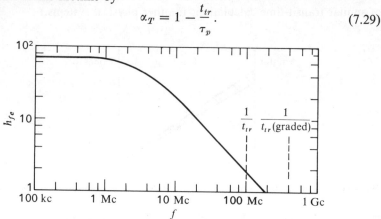

Fig. 7.12 The effect of signal frequency on the common-emitter current gain of the *pnp* transistor. ($I_C = 50$ ma, $V_{CE} = -3$ v)

Physically, this is very suggestive—the probability of recombination of an injected hole in the base region will indeed depend on this ratio.

b. Graded Base Regions

So far we have based our discussion on the idealized transistor model in which the distribution of the impurities in the base region was assumed to be uniform. In double-diffused transistors, the distribution of impurities in the base region is not uniform, but is quite strongly graded. The estimated distribution of impurities in our *pnp* transistor is shown in Figure 7.13. Note the very significant variation of impurity concentration across the base region. This impurity concentration profile will bring about a similar variation of majority carrier concentration in the base region. However, in equilibrium no current flows so that an electric field must exist in the neutral base region. This electric field will counterbalance the diffusion current due to the concentration gradient of majority carriers existing there. (This is also evident from the corresponding band diagram shown in Figure 7.14.) If minority carriers are now injected into the base, their motion will be affected by the finite electric field which is present in the neutral base region.

Limitations and Modifications of the Simple Theory

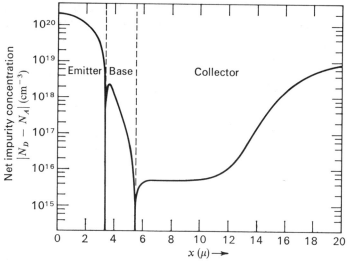

Fig. 7.13 The impurity distribution in the *pnp* transistor. (Estimated on the basis of diffusion conditions.)

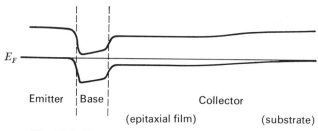

Fig. 7.14 The corresponding band diagram, in equilibrium.

In the case of *pnp* transistors, because of the impurity concentration gradient, the electrons within the base tend to diffuse toward the collector. Thus an electric field must be present, pushing the electrons toward the emitter-base junction. This same electric field will then be of such direction as to aid the motion of injected holes. Thus *the injected minority carriers will now move not only by diffusion but also by drift* due to the existence of this electric field. As a result, the transit time across the base will decrease and the upper frequency limitation of the transistor associated with this transit time will increase.[2] Correspondingly, the transport factor will also increase.

To provide an indication of the effect of this built-in electric field, in Figure 7.15 theoretical calculations[3] of the transit time are shown as a function of the ratio of the concentration of the base impurity at the emitter-base junction to that at the collector-base junction. In these calculations, the impurity distribution in the base region was approximated by an erfc-type distribution. The transit time for the *pnp* transistor, corrected by using Figure 7.15, is also indicated in Figure 7.12.

The calculations of Moll and Ross,[3] on which Figure 7.15 was based, also show that in the derivation of the currents flowing in a transistor, the quantity $(N_{DB}W_B)$ in the denominator of Equation 7.12 and in subsequent

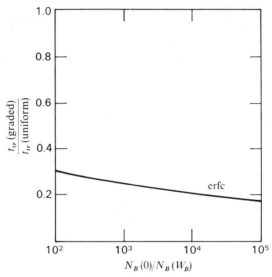

Fig. 7.15 Reduction in base transit time due to the graded impurity concentration across the base region. $N_B(0)$ and $N_B(W_B)$ are the base-impurity concentrations at the emitter-base and collector-base junctions, respectively.[3]

equations is replaced by the total number of impurities in the base per unit area Q_B, given by

$$Q_B \equiv \int_0^{W_B} N_{DB}(x)\, dx. \tag{7.30}$$

Furthermore, N_{DB} itself is replaced by its value at the emitter-base junction, $N_{DB}(0)$. Otherwise, the equations remain unchanged. With these changes, results based on the idealized transistor model can be readily adapted to the consideration of double-diffused transistors.

c. Early Effect

The effect of the collector-base reverse bias on the common emitter current gain h_{FE} is not explicit in the equations of the previous sections.

However, it is evident that as the collector-base junction is reverse biased, the width of the collector-base depletion region will increase and consequently *the width of the neutral base region W_B will be reduced.* Thus the gradient of the injected minority carriers in the neutral base region will become steeper, and therefore the collector current will increase. The base current, however, will not change significantly since it is primarily due to phenomena occurring near the emitter-base junction.

Thus as the collector-base junction reverse bias is increased, the current gain will increase. This phenomenon, first pointed out by Early[4] and since then commonly referred to as the *Early effect*, is clearly evidenced in the characteristics of Figure 7.4b. These characteristics are traced under constant base current conditions. It is evident that as the collector-to-emitter bias and therefore the collector-base reverse bias is increased, the collector current for a given base current increases as represented by the upward tilting of the characteristics.

The Early effect is, of course, much more pronounced if the doping concentration in the base region is relatively light as compared to the doping concentration in the collector region.

d. High Current Effects

The simple theory developed in Section 7.2 predicts that, as the current level is increased, h_{FE} should reach a constant, current-independent level. This theory does not consider deviations from the condition of low-level injection.

Actually, at high current levels, the injected carrier concentration may greatly exceed the doping concentration in the base region, i.e., $p_n \gg N_{DB}$ may hold. In such a case, in order to maintain charge neutrality within the base, the electron and hole concentrations there must become equal. Thus *the majority carrier concentration will also increase as the minority carrier concentration is increased.* Such condition is referred to as *modulation of the conductivity* of the semiconductor.

In the conductivity-modulated condition the semiconductor in effect becomes more heavily "doped" as the injection level is increased. As a result, the rate of increase of the injection level with increasing forward bias slows down. Thus the collector current will no longer follow the simple exponential law,

$$I_C \propto e^{qV_{EB}/kT},$$

but will approach[5]

$$I_C \propto e^{qV_{EB}/2kT}.$$

The slower increase of injected current with forward bias at high current levels is indeed evident in Figure 7.9. It is also evident in the forward current-voltage characteristics of the germanium, silicon, and gallium arsenide diodes shown in Figure 6.24.

Corresponding to the increase in majority carrier concentration in the base region, the emitter efficiency will decrease bringing about a reduction in current gain at high current levels.[6] A reduction in current gain is in fact observed at high collector currents in Figure 7.10.

Another important deviation from the low-level theory is due to the fact that the simple theory of p-n junctions which we have discussed in Chapter 6 and on which the above treatment of transistors has been based is founded on the basic underlying assumption that the semiconductor is divided into depletion regions in which the carrier concentrations are smaller than the impurity concentration, and into neutral regions where space-charge neutrality approximately prevails. At high current levels, significant space charge may be present everywhere in the semiconductor, wiping out a meaningful distinction between depletion regions and neutral regions. Thus, the analysis of semiconductor devices must be modified for such a condition.[7] This is beyond the scope of the present discussion.

7.4 BASE RESISTANCE

We have seen that due (i) to injection of minority carriers into the emitter region, (ii) to recombination in the emitter-base space-charge region, and (iii) to recombination in the base region a current will flow to the base lead. This current flows in a direction transverse to the direction of the normal transistor current flow as shown in Figure 7.16. As a result, a voltage drop will build up in the base region along the path of the base current flow. In the case of the *pnp* transistor shown here, the potential will be highest at

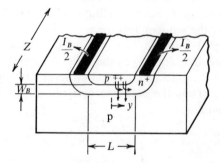

Fig. 7.16 Illustration of the calculation of the base spreading resistance for a stripe geometry.

Base Resistance

the center line under the emitter region, and lowest near the base contact. This voltage build-up is important because it results in a variation of the emitter-base forward bias as a function of distance y, leading to a *higher forward bias near the edges* of the emitter region than in the central section. This, in turn, results in a higher current density near the edges of the emitter. This condition is referred to as *current crowding*. Thus we want both to calculate and to control the *base spreading resistance* through which the base current must flow.

The average magnitude of the base voltage drop is given by

$$\bar{V}_B = \frac{1}{\tfrac{1}{2}L} \int_0^{\tfrac{1}{2}L} V_B(y)\, dy \tag{7.31}$$

if the voltage drop outside the narrow region between the emitter and the collector can be neglected. Then the base spreading resistance can be defined by

$$r_B' \equiv \frac{\bar{V}_B}{I_B}. \tag{7.32}$$

The calculation of the average base voltage drop $\bar{V}_B$ is dependent on the particular transistor geometry. Such a calculation can be illustrated by considering the simple stripe geometry shown in Figure 7.16. The voltage drop over an element dy here is given by

$$-dV_B = \frac{\bar{\rho}_B\, dy}{Z W_B} I_B(y) \tag{7.33}$$

where Z is the length of the stripe, and $\bar{\rho}_B$ is the average resistivity of the base region. The base current at point y is assumed to be given by

$$I_B(y) = \tfrac{1}{2} I_B \frac{y}{\tfrac{1}{2}L}. \tag{7.34}$$

Here I_B is the total base current. (This assumes no current crowding.) Substitution of Equation 7.34 into 7.33 and integration leads to

$$\bar{V}_B = \frac{1}{12} \frac{\bar{\rho}_B L}{W_B Z} I_B. \tag{7.35}$$

The base spreading resistance is obtained by dividing the voltage drop by I_B, and is

$$r_B' = \frac{1}{12} \frac{\bar{\rho}_B}{W_B} \frac{L}{Z}. \tag{7.36}$$

The reciprocal of the average resistivity of the base region is given by

$$\frac{1}{\bar{\rho}_B} = \frac{1}{W_B} q \int_0^{W_B} \mu_n [N_D(x) - N_A(x)] \, dx. \tag{7.37}$$

A quick estimate of the average resistivity can be obtained by multiplying the average base impurity concentration Q_B/W_B with the mobility corresponding to this concentration. Thus,

$$\frac{1}{\bar{\rho}_B} \approx q\mu_n \frac{Q_B}{W_B} \tag{7.38}$$

where μ_n is the mobility value corresponding to the impurity concentration Q_B/W_B.

7.5 MAXIMUM VOLTAGE LIMITATIONS

a. Common Base Configuration

The maximum voltage that can be applied to a transistor in the common-base configuration, BV_{CBO} (see Figure 7.4a), is usually determined by the *avalanche breakdown voltage* of the collector-base junction.

Avalanche breakdown in *p-n* junctions has been discussed in the previous chapter. That discussion applies without modification to the common-base breakdown in transistors in such a case.

b. Common-Emitter Configuration

The common-emitter breakdown voltage BV_{CES} is measured with the base shorted to the emitter during the measurement. The breakdown observed under this condition will usually be the avalanche breakdown of the collector-base junction. In that case, $BV_{CES} = BV_{CBO}$. However, an interesting departure from this can take place if the space-charge region on the base side of the collector-base junction reaches the emitter-base junction before the collector-base junction can avalanche. This situation, which is referred to as the *punch-through* condition, is illustrated in Figure 7.17. Once the collector-base depletion region reaches the emitter-base junction, the two *p*-regions—the emitter and the collector—are connected with a continuous depletion region. A current can now flow, hence "breakdown" takes place even in the absence of any avalanche process.

To calculate the voltage required to bring about the punch-through

Maximum Voltage Limitations

condition, we estimate the area under the electric field distribution curve shown in Figure 7.17. This yields

$$BV = \frac{qQ_B}{K_s\epsilon_o}\left[W_B + \frac{Q_B}{2N_A}\right] \quad (7.39)$$

where Q_B is the total number of impurities per unit area in the base region and N_A is the acceptor concentration in the collector. Because the base

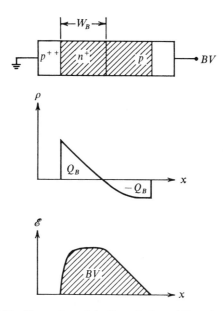

Fig. 7.17 Illustration of the "punch-through" condition.

region of diffused transistors is relatively heavily doped, the punch-through voltage is usually high in diffused transistors, and therefore the voltage limitation is usually due to the avalanche breakdown of the collector-base junction.

The maximum voltage in the common-emitter configuration with the base lead open, BV_{CEO}, is of particular importance (see Figure 7.4b). To calculate this voltage, we must first consider the currents flowing in the transistor in such a two-terminal operation. When a certain voltage V_{CE} is applied to the collector with respect to the emitter, with the base floating, as shown in Figure 7.18, the base region will acquire a potential that is intermediate between the emitter potential and the collector potential. As a result, *the emitter-base junction will be very slightly forward biased.* In such a case, the collector current will consist of the reverse-biased generation current of the collector-base junction, I_{gen}, *plus* the current carried by those injected carriers which reach the collector-base junction. If the emitter current is I_E, the latter component will be given by $\gamma \alpha_T I_E$.

However, the current flowing through the emitter-base junction and the collector-base junction must be the same. Hence, in the open-base configuration,

$$I_E = I_C = \gamma \alpha_T I_E + I_{\text{gen}}. \tag{7.40}$$

The generation current is just the leakage current of the collector-base junction measured under open emitter conditions I_{CBO}. Thus the above relation leads to

$$I_{CEO} = \gamma \alpha_T I_{CEO} + I_{CBO}. \tag{7.41}$$

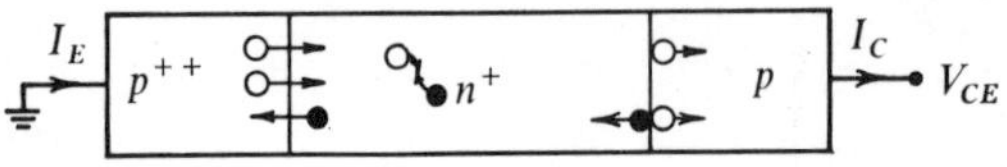

Fig. 7.18 Illustration of current flow in the open-base condition.

So far we have neglected the role of multiplication in the collector-base junction. It was pointed out in Chapter 6 that near breakdown the current normally flowing in a reverse biased *p-n* junction is multiplied by the factor M because of the incipient avalanche process. Thus, near breakdown, the above relationship will lead to

$$I_{CEO} = (\gamma \alpha_T I_{CEO} + I_{CBO})M \tag{7.42}$$

or

$$I_{CEO} = \frac{I_{CBO} M}{1 - \gamma \alpha_T M}. \tag{7.43}$$

This relationship shows that the leakage current in the common-emitter configuration is larger than the leakage current in the common-base configuration by the factor $1/(1 - \gamma \alpha_T M)$. As a result, this current will start increasing rapidly when $\gamma \alpha_T M \to 1$, rather than when $M \to \infty$ as was the case in the common-base configuration. Thus the *breakdown voltage in the common-emitter configuration will be lower.*

To estimate the reduction in breakdown voltage, we can use the empirical formula for the multiplication factor discussed in Chapter 6,

$$M = \frac{1}{1 - \left(\dfrac{V_{CB}}{BV_{CBO}}\right)^n} \tag{7.44}$$

where BV_{CBO} is the true breakdown voltage of the collector-base junction, and n is between 3 and 6.

Maximum Voltage Limitations

Substituting Equation 7.44 into the condition, $\gamma \alpha_T M = 1$ when $V_{CB} = BV_{CEO}$, leads to the relationship,

$$\frac{BV_{CEO}}{BV_{CBO}} = \sqrt[n]{1 - \gamma \alpha_T} \simeq \frac{1}{\sqrt[n]{h_{FE}}}. \tag{7.45}$$

We have seen in Chapter 6 that the breakdown voltage of *planar* junctions is determined by the electric field in the rounded, nearly cylindrical

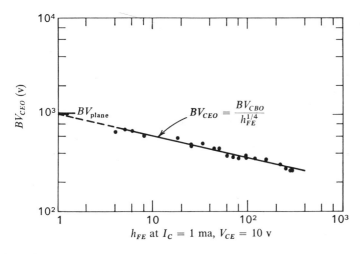

Fig. 7.19 Open-base breakdown voltage of *npn* transistors as a function of common-emitter current gain.[8]

region of the junction. In Chapter 10 we will see that charges near the surface can increase the electric field and further lower the breakdown voltage of the junction. Thus it is important to consider which if any of these factors will also have an effect on the common-emitter breakdown voltage BV_{CEO}. The above discussion applies only to the region of the transistor under the emitter, i.e., to the section indicated in Figure 7.1a, because injection in other regions of the transistor, e.g., near the surface, or near the corner region, is much less efficient. Thus the correct breakdown voltage BV_{CBO} that must be used in Equation 7.45 is that of a truly *plane* junction which is often higher than the actual BV_{CBO} of a planar transistor.

This is illustrated by the experimental measurements[8] shown in Figure 7.19, which show the breakdown voltage of planar *npn* transistors in the common-emitter mode. Note that the power-law relationship to the current gain is observed and that the straight line extrapolates to the true plane breakdown-voltage value. Similar results were obtained with *pnp* transistors, yielding $n \simeq 6$.

It is interesting to point out that since the current gain h_{FE} itself is a function of collector current, the reduction in breakdown voltage as given by Equation 7.45 will also depend on the collector current level. Thus the breakdown characteristic in the common-emitter mode will exhibit a switchback as shown in the example in Figure 7.4b due to the variation of h_{FE} as a function of collector current.

7.6 MINIMUM VOLTAGE LIMITATIONS

When a transistor is operated in the common-base configuration, the same collector current will flow when there is zero bias applied across the collector-base junction as when this junction is reverse biased. Thus in this mode of operation the transistor imposes no lower limit upon the output voltage. The situation, however, is different in the common-emitter configuration.

A constant base current approximately corresponds to a constant emitter-base forward bias, e.g., 0.5 v. As the collector-to-emitter voltage is now reduced to this value, the collector-base junction cannot be reverse biased any more. Further reduction in V_{CE} will actually cause the collector-base junction to be *forward biased*, injecting carriers in the opposite direction to those injected from the emitter. Thus *two forward-biased p-n junctions now face each other*. As a result, the measured net collector current, which is the difference of the two injected currents, will decrease as V_{CE} is decreased as shown in Figure 7.20. This condition is called *saturation*.

Saturation was considered by Ebers and Moll[9] who treated the two forward-biased junctions independently and calculated the net collector current flowing through the transistor by superposition. The individual and net carrier distributions due to forward-biasing of just one and of both junctions is illustrated in Figure 7.21. Note the reduction in collector current as evidenced by the decreased magnitude of the slope of the net minority carrier distribution in Figure 7.21b.

To simplify the analysis, the recombination current components associated with the two depletion regions are neglected and the transport factor α_T is taken to be unity. The first assumption is justified at relatively high current levels; the second focuses attention on the effect of the different emitter efficiencies of the two junctions. This effect is particularly important in double-diffused transistors where injection from the collector to the base is an inefficient process due to the higher doping concentration

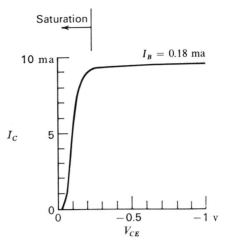

Fig. 7.20 Collector current of the *pnp* transistor in saturation.

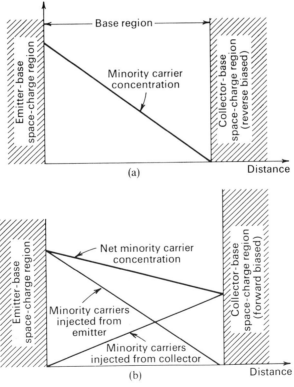

Fig. 7.21 The minority carrier distribution within the base region.
(a) Emitter-base junction forward biased, collector-base junction reverse biased.
(b) Both junctions forward biased: saturation.

in the base. Finally, equal junction areas are assumed. The result,[9] after lengthy algebra, is

$$V_{CE} = \pm \left\{ \frac{kT}{q} \left| \ln \frac{\alpha_R \left[1 - \frac{I_C}{I_B h_{FE}}\right]}{1 + \frac{I_C(1-\alpha_R)}{I_B}} \right| + |I_E r_{SE}| + |I_C r_{SC}| \right\} \quad (7.46)$$

where r_{SE} and r_{SC} are the series resistances of the emitter and the collector regions, and α_R is the common-base current gain in the *reverse direction*, i.e., using the collector as the emitting junction. (α_R is generally much smaller than α for double-diffused transistors.) The positive sign is taken for *npn* transistors, the negative for *pnp*.

Note the appearance of the term $I_C/(I_B h_{FE})$ in Equation 7.46. Out of saturation, where the current gain h_{FE} is defined, this term is unity. Thus the argument of the logarithm is zero and therefore $V_{CE} \to -\infty$ as it indeed does on the characteristics (see Figure 7.20). In saturation, however, this term is no longer unity, thereby leading to a finite value of V_{CE}—the minimum voltage at which the transistor can be operated in the common-emitter mode.

7.7 THERMAL LIMITATION

Carriers accelerated in the large electric field of the collector-base depletion region (see Figure 7.2b) suffer many collisions with the semiconductor lattice, thereby converting their kinetic energy into heat. The power dissipated this way, $I_C V_{CB}$, must be transported away from the collector-base junction if the temperature of the junction is to remain at a steady-state value. The path through which this heat must be conducted away is illustrated schematically in Figure 7.22. First, the heat is transported from the junction, which has a temperature T_J, to the supporting

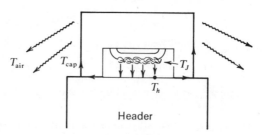

Fig. 7.22 Illustration of the flow of heat in an encapsulated transistor.

Thermal Limitation

header through the silicon. From the header directly under the junction, which has a temperature designated by T_h, the heat flows through the header material to other parts of the package, e.g., the cap, which are directly in contact with the surrounding ambient. The cap will have a temperature T_{cap} which is lower than T_h. Finally, the heat will be conducted away from the transistor package to the surrounding ambient.

The total resistance to the flow of heat can be pictured as a series combination of the individual resistances of the silicon, of the header, and of the air (in the absence of a heat sink), as illustrated by the simple circuit analogue shown in Figure 7.23. In this circuit, we also show what happens

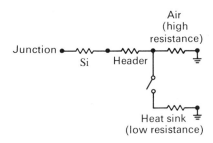

Fig. 7.23 Circuit analogue for the heat-flow problem.

if the transistor package is placed in contact with a heat sink, such as a metal block. A heat sink will provide a small shunt resistor in parallel to the resistance due to the air.

Let us consider the individual resistances appearing in the case of an air-cooled device. The resistance for thermal flow is obtained from Fourier's law of heat conduction,

$$P = -k_{th} \frac{\partial T}{\partial x} A \qquad (7.47)$$

which states that the heat flow rate P is proportional to the temperature gradient.† The proportionality constant k_{th} is the thermal conductivity; A is the cross-sectional area available for the flow of heat. In steady state this becomes

$$P = k_{th} \frac{\Delta T}{L} A \qquad (7.48)$$

where L is the length of the path of heat conduction. This can be recognized as the equivalent of Ohm's law for the conduction of heat. Thus the

† Note the analogy to diffusion where the flux is proportional to the concentration gradient.

thermal resistance of a material will be given by

$$R_{th} = \frac{L}{k_{th}A}. \quad (7.49)$$

Accordingly, the individual resistances that we must combine in series are

$$R_{th,\mathrm{Si}} = \frac{L_{\mathrm{Si}}}{k_{th,\mathrm{Si}}A_{\mathrm{Si}}} \quad \text{for the silicon} \quad (7.50)$$

$$R_{th,h} = \frac{L_h}{k_{th,h}A_h} \quad \text{for the header} \quad (7.51)$$

$$R_{th,\mathrm{air}} = \frac{1}{h_{th}A_{\mathrm{air}}} \quad \text{for the air} \quad (7.52)$$

where h_{th} is the gas-phase heat-transfer coefficient, analogous to the mass-transfer coefficient discussed in Chapter 1. Typical values for the quantities involved here are:

$L_{\mathrm{Si}} \simeq$ thickness of wafer $\simeq 10^{-2}$ cm,

$A_{\mathrm{Si}} \simeq$ junction area, $\sim 10^{-3}$ cm² for a medium-sized device,

$k_{th,\mathrm{Si}} \simeq 1.5$ watt/(°C cm)

hence, $R_{th,\mathrm{Si}} \simeq 5$°C/watt;

$L_h \simeq 1$ cm,

$A_h \simeq 10^{-2}$ cm²,

$k_{th,h} \simeq 4$ watt/(°C cm)

hence, $R_{th,h} \simeq 5$°C/watt;

$h_{th} \simeq 10^{-3}$ to 10^{-2} watt/(°C cm²),

$A_{\mathrm{air}} \simeq 1$ cm²

hence, $R_{th,\mathrm{air}} \simeq 10^2$ to 10^3 °C/watt.

Thus the overall thermal resistance will be 10^2 to 10^3 °C/watt if the device is air-cooled. This resistance can be reduced to about 10°C/watt by proper heat sinking.

READING REFERENCES

Junction transistors are the subject of several texts. Among these are:

R. D. Middlebrook, *An Introduction to Junction Transistor Theory*, Wiley, 1957.

W. W. Gärtner, *Transistors*, Van Nostrand, 1960;

A. B. Phillips, *Transistor Engineering*, McGraw-Hill Book Co., 1962;

S. Hakim, *Junction Transistor Circuit Analysis*, Wiley, 1962.

REFERENCES CITED

1. Transistor action was first described by J. Bardeen and W. H. Brattain, "The Transistor, A Semiconductor Triode," *Phys. Rev.*, **74**, 230 (1948). The theory of current flow in junction transistors was established by W. Shockley, "The Theory of *P-N* Junctions in Semiconductors and *P-N* Junction Transistors," *Bell System Tech. J.*, **28**, 435 (1949). This theory was then extended by C. T. Sah, R. N. Noyce, and W. Shockley, "Carrier Generation and Recombination in *P-N* Junctions and *P-N* Junction Characteristics," *Proc. IRE*, **45**, 1228 (1957).
2. H. Kroemer, "Der Drifttransistor," *Naturwiss*, **40**, 578 (1953).
3. J. L. Moll and I. M. Ross, "The Dependence of Transistor Parameters on the Distribution of Base Layer Resistivity," *Proc. IRE*, **44**, 72 (1956).
4. J. M. Early, "Effects of Space-Charge Layer Widening in Junction Transistors," *Proc. IRE*, **40**, 1401 (1952).
5. R. N. Hall, "Power Rectifiers and Transistors," *Proc. IRE*, **40**, 1512 (1952).
6. W. M. Webster, "On the Variation of Junction-Transistor Current Amplification Factor with Emitter Current," *Proc. IRE*, **42**, 914 (1954).
7. See, for instance, C. T. Kirk, "A Theory of Transistor Cutoff Frequency (f_T) Falloff at High Current Densities," *IRE Trans.*, *Electron Devices* **ED-9**, 164 (1962); A. van der Ziel and D. Agouridis, "The Cutoff Frequency Falloff in UHF Transistors at High Currents," *Proc. IEEE* (Correspondence), **54**, 412 (1966).
8. W. W. Hooper, unpublished.
9. J. J. Ebers and J. L. Moll, "Large Signal Behavior of Junction Transistors," *Proc. IRE*, **42**, 1761 (1954).

PROBLEMS

7.1 Derive the expression giving the small-signal current gain h_{fe}, Equation 7.6.

7.2 (a) Verify that Equation 7.10 satisfies the transport equation and the boundary conditions.
(b) Verify the exact and approximate expressions given for α_T.

7.3 If h_{FE} for the case when $\alpha_T = 1$ is denoted by $h_{FE,\gamma}$, and h_{FE} for the case when $\gamma = 1$ is denoted by h_{FE,α_T}, show that, in general,

$$\frac{1}{h_{FE}} = \frac{1}{h_{FE,\gamma}} + \frac{1}{h_{FE,\alpha_T}}.$$

7.4 When a transistor is irradiated with large doses of fast neutrons or high-energy electrons, it is often found that the lifetime decreases with irradiation dose approximately as $\tau = K/\phi$, where ϕ is the dose and K is an empirical constant (see Chapter 5). Assuming that the emitter efficiency is unity, derive a relationship giving (a) base current, and (b) h_{FE} as a function of the dose, and their respective values before irradiation.

7.5 If the total charge due to injected excess minority carriers within the base is Q, and if the emitter efficiency $\gamma = 1$, express the base current, collector current, and the current gain in terms of Q.

7.6 (a) Estimate the magnitude of the electric field near the emitter-base junction, in equilibrium, for the *pnp* transistor used for illustration in this chapter. (*Hint*: Fit an exponential profile to the base impurity distribution shown in Figure 7.13.)
(b) Calculate the fraction of the injected minority-carrier flow that is due to drift when the emitter-base junction is forward biased. Assume that the electric field is unchanged. (Under what conditions is this assumption justified?)
(c) Recalling that a flux is the product of concentration and velocity, use your result to estimate the increase of the velocity of the injected minority carriers due to the built-in field. Compare your result with Figure 7.15.

7.7 (a) Using the collector current versus emitter-base voltage data for the *pnp* transistor, and assuming a lifetime of 1 μsec, calculate the total number of impurities per unit area in the base.
(b) Using the maximum value of the current gain, calculate the total number of impurities per unit area in the emitter.
Compare both numbers with the corresponding estimates based on the impurity concentration distribution given.

7.8 Derive an expression showing the effect of the collector-base junction reverse bias on the transit time across the base and on h_{FE}, assuming that the collector-base junction is linearly graded.

7.9 The collector current is transported across the reverse-biased depletion region of the collector-base junction by drift.
(a) Assuming that the velocity of the carriers is their maximum drift velocity, show that the concentration of injected carriers across the base-collector depletion region is constant.
(b) Sketch the electric field distribution within the collector-base junction depletion region for increasing current densities, assuming the base is much more heavily doped than the collector.
(c) At what current density does the electric field approach a constant value?
(d) What will happen if the current density is further increased?

7.10 Calculate the current density at which the base becomes conductivity modulated for a *pnp* transistor. Compare this current density with that obtained in Problem 7.9c. Which condition sets in first in a double-diffused transistor? In an alloy transistor?

7.11 One criterion of the onset of current crowding is when the transverse base voltage drop exceeds kT/q. Estimate the corresponding collector current level for a *pnp* transistor which has an h_{FE} of 50, an impurity distribution as given in Figure 7.13, and a stripe geometry with $Z = 40$ mil, $L = 0.5$ mil.

7.12 In a transistor in which the value of BV_{CEO} is determined by "punch-through", $|BV_{CBO}| = |BV_{CEO}| + |BV_{EBO}|$. Justify this relationship. (*Hint*: Consider what happens at the emitter-base junction as "punch-through" is obtained.)

7.13 A reverse bias V_R is applied to the base region with respect to the emitter. What effect will this bias have on the voltage applied to the collector, also with respect to the emitter, at which large reverse current begins to flow? Consider both a

Problems

transistor which is avalanche-breakdown limited and one which is punch-through limited. Can this test be used to distinguish the two cases?

7.14 For transistor "A",

$BV_{CBO} = 105$ v.
$BV_{CES} = 105$ v.
$BV_{CEO} = 96$ v.
$BV_{EBO} = 9$ v.

For transistor "B",

$BV_{CBO} = 75$ v.
$BV_{CES} = 69$ v.
$BV_{CEO} = 69$ v.
$BV_{EBO} = 6$ v.

What mechanism limits the maximum voltage of these two transistors?

7.15 Draw the energy band diagram for a *pnp* transistor in saturation.

7.16 The Ebers-Moll analysis makes three important simplifying assumptions: It neglects the recombination current components associated with the two depletion regions; it assumes that both transport factors are unity; and it assumes that the areas of the emitter and collector junctions are equal. Discuss the validity and importance of these assumptions, and indicate qualitatively the direction of any errors they may introduce with respect to:
(a) Alloy germanium transistors.
(b) Planar, double-diffused silicon transistors.

7.17 Calculate the overall thermal resistance of the device discussed in Section 7.7, also taking into account conduction of heat through the leads. Assume a transistor, with a copper base and a copper emitter lead, each 2 mil in diameter, and 5 mm long. Do you expect the two leads to be equally effective as heat-conduction paths? Justify.

7.18 A single pulse of current is imposed upon a silicon transistor of dimensions as given in Section 7.7. How long a duration must the current pulse have for the transistor chip to achieve its new steady-state temperature? In this calculation neglect heat conduction away from the chip.

TABLE 7.1
IMPORTANT FORMULAS FOR JUNCTION TRANSISTORS

	pnp	npn
Current gain	$\alpha \equiv h_{FB} \equiv \dfrac{I_C}{I_E} \qquad \beta \equiv h_{FE} \equiv \dfrac{I_C}{I_B} \qquad \beta = \dfrac{\alpha}{1-\alpha}$ $\alpha = \gamma \alpha_T$ [See footnote on page 219.]	
Transport factor	$\alpha_T \cong 1 - \dfrac{1}{2}\left(\dfrac{W_B}{L_{pB}}\right)^2$ $\cong 1 - \dfrac{t_{tr}}{\tau_p}$	$\alpha_T \cong 1 - \dfrac{1}{2}\left(\dfrac{W_B}{L_{nB}}\right)^2$ $\cong 1 - \dfrac{t_{tr}}{\tau_n}$
Emitter efficiency	$\gamma = \dfrac{1}{1 + \dfrac{B}{E} + \dfrac{1}{2}\sqrt{\dfrac{qBA_J}{I_C}}\,R}$ $R = \dfrac{W_{EB}}{\tau_o} + s_o \dfrac{A_s}{A_J}$ $B \equiv \dfrac{N_{DB}W_B}{D_{pB}}$ $E \equiv \dfrac{N_{AE}W_E}{D_{nE}}$	$B \equiv \dfrac{N_{AB}W_B}{D_{nB}}$ $E \equiv \dfrac{N_{DE}W_E}{D_{pE}}$
Transit time	$t_{tr} = \dfrac{W_B{}^2}{2D_{pB}}\,;$	$t_{tr} = \dfrac{W_B{}^2}{2D_{nB}}$
Base resistance	$r_B{}' = \dfrac{1}{12}\dfrac{\bar{\rho}_B}{W_B}\dfrac{L}{Z}$ for stripe geometry	
Leakage currents	$I_{CEO} = \dfrac{I_{CBO}\,M}{1 - \gamma \alpha_T M}$	
Maximum voltages	$BV_{CEO} \cong \dfrac{BV_{CBO}}{\sqrt[n]{h_{FE}}}$	
Minimum voltage	$V_{CE}(\text{sat}) = \pm \left\{ \dfrac{kT}{q}\left|\ln \dfrac{\alpha_R\left[1 - \dfrac{I_C}{I_B h_{FE}}\right]}{1 + \dfrac{I_C(1 - \alpha_R)}{I_B}}\right| + |I_E r_{SE}| + |I_C r_{SC}| \right\}$ $-\text{pnp} \qquad\qquad +\text{npn}$	

- **PRINCIPLES OF OPERATION**
- **CHARACTERISTICS**
- **MODIFICATIONS OF THE SIMPLE THEORY**

8
Junction Field-Effect Transistors

The junction field-effect transistor, often called field-effect transistor, proposed by Shockley[1] in 1952 and first demonstrated by Dacey and Ross,[2] is a device based on an entirely different physical principle than the junction transistor. While the junction transistor operates through the transport of injected minority carriers, in a junction field-effect transistor the depletion region of reverse-biased *p-n* junctions is used to modulate the cross-sectional area available for current flow. The current is transported by carriers of one polarity only; hence, it is usual to refer to the field-effect transistor as a *unipolar* device in contrast to the junction transistor which is a *bipolar* device since it involves both types of carriers.

The field-effect transistor in its most common form is illustrated in Figure 8.1. An *n*-type layer is grown epitaxially on a heavily doped *p*-type substrate. Then, using regular planar technology, a *p*-type *gate* junction is formed by diffusion from the top. Finally, contact is made to the *p*-regions and also to the *n*-region at either side of the top gate resulting in a *source* and a *drain* contact.† With such fabrication techniques, the top and bottom gate junctions are approximately symmetrical. Accordingly, in the following discussion we concentrate our attention on symmetrical field-effect transistors.

† In order to insure good contact to the *n*-region, it is usually necessary to increase its surface concentration. This can be accomplished by an additional *n*-type diffusion at the source and drain regions. These are omitted in Figure 8.1.

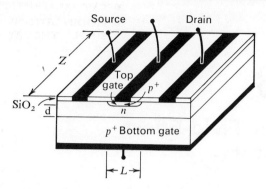

Fig. 8.1 n-Channel junction field-effect transistor fabricated by planar-epitaxial methods.

We begin by discussing the physical principles underlying the operation of the junction field-effect transistor. Then we derive the most important characteristics of such devices, including the current-voltage relationship, the channel conductance, and the transconductance, and discuss the factors affecting the gate leakage current. Finally, we discuss the various modifications of the simple theory of field-effect transistors.

8.1 PRINCIPLES OF OPERATION

In Figure 8.2, we illustrate the conditions that prevail when the gate-to-source potential $V_G = 0$. If a small positive voltage V_D is applied to the drain, electrons will flow from source to drain; hence, a current will flow from drain to source through the n-type region enclosed between the two depletion regions. Such a region is commonly referred to as a *channel*

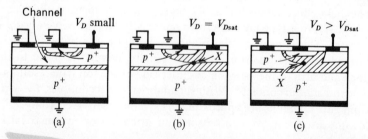

Fig. 8.2 Illustration of the operation of a junction field-effect transistor for $V_G = 0$.
(a) V_D is small; channel resistance is constant.
(b) $V_D = V_{D\,\text{sat}}$; onset of saturation.
(c) $V_D > V_{D\,\text{sat}}$; no further increase in drain current.

Principles of Operation

of n-type material. Accordingly, the device shown in Figures 8.1 and 8.2 is referred to as an n-channel junction field-effect transistor.

The resistance of the channel is given by

$$R = \frac{L}{q\mu_n N_D Z(d - 2W)} \tag{8.1}$$

where N_D is the donor concentration in the channel region, L, Z, and d designate the length, width, and thickness of the channel, respectively (see Figure 8.1), and W is the width of the depletion region of the top and bottom gates.

In the case shown in Figure 8.2a, when no gate voltage is applied and V_D is small, W is the zero-bias depletion region of the gate junctions. It is evident from Figure 8.2a that these depletion regions restrict the current flow to a smaller cross-sectional area than would exist without them.

For any given drain voltage, the voltage along the channel will increase from zero at the source to V_D at the drain. Thus both gate junctions will become increasingly reverse biased as we proceed from the source to the drain. So long as V_D is much less than the built-in voltage of the gate junctions ϕ_B, the depletion region width W will remain practically independent of V_D and the channel will act as a resistor. However, as V_D is increased, the average cross-sectional area for current flow is reduced because of the increasing reverse bias of the gate junctions near the drain area. Thus the channel resistance R will also increase. As a result, the current-voltage characteristics will begin to fall below the initial resistor line. This tendency is evident in the experimental measurements shown in Figure 8.3 where drain current is plotted as a function of drain voltage for a silicon junction field-effect transistor, used for illustration throughout this chapter. The top curve is for $V_G = 0$.

As the drain voltage V_D is further increased, the depletion region width also increases near the drain until eventually the two depletion regions touch, as indicated in Figure 8.2b. This happens when

$$W = \frac{d}{2}. \tag{8.2}$$

Using the one-sided step-function formulas, we can readily obtain the corresponding value of the drain voltage V_{Dsat},

$$V_{Dsat} = \frac{qN_D d^2}{8K_s\epsilon_0} - \phi_B \quad [V_G = 0] \tag{8.3}$$

where ϕ_B is the built-in voltage of the gate junctions.

At this drain voltage, the source and the drain are completely separated by a reverse-biased depletion region which normally does not conduct because there are very few carriers in it. However, we have already seen in Chapter 6 that if carriers are created within a reverse-biased depletion region by thermal generation or by avalanche breakdown, a current will flow across it. We have also seen in Chapter 7 that large currents can flow across a reverse-biased depletion region if carriers are injected into it—as

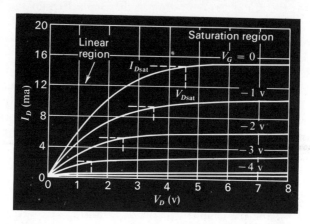

Fig. 8.3 Current-voltage characteristics of a silicon *n*-channel junction field-effect transistor. This device is used for illustration throughout this chapter. Its structural parameters are: $Z/L = 170$, $d = 3\ \mu$, $N_D = 2.5 \times 10^{15}$ cm^{-3}.

in the collector-base depletion region of a transistor when the emitter-base junction injects minority carriers into it, or in the case of punch-through breakdown.

The situation existing in a field-effect transistor after the depletion regions have met is quite similar. A current will flow in this case across the combined depletion regions separating the source from the drain. This current flows because of carriers injected into the depletion region from the channel at the point where the depletion regions touch, marked by X in Figure 8.2b. As in the case of the current flowing in the depletion region of the collector-base junction of a transistor, the current flowing across the depletion region beyond the point X will be limited by the number of carriers arriving at this point. The magnitude of this current, in turn, depends on the voltage drop from the source to the point X. *This voltage drop, however, is exactly* $V_{D\text{sat}}$ *since this is the reverse bias that is required for the two depletion regions to touch.*

If the drain voltage V_D is increased beyond $V_{D\text{sat}}$, the depletion region near the drain will merely thicken and the point X will move slightly toward the source, as indicated in Figure 8.2c. However, the voltage at

Principles of Operation

the point X will still remain the same, $V_{D\text{sat}}$. Thus the number of carriers arriving from the source to this point, and hence the current flowing from source to drain, will remain unaltered since the potential drop in the channel from source to the point X remains unaltered.† Thus *for drain voltages larger than $V_{D\text{sat}}$ the current will not change but will remain at the value $I_{D\text{sat}}$*, as is evident from the experimental data shown in Figure 8.3. This phenomenon is called *saturation* because the current saturates with increasing drain voltage.

When a gate voltage is applied to both *p*-regions of such a polarity as to reverse bias the gate-to-channel junctions (i.e., negative for an *n*-channel device), the depletion regions will, of course, become wider.‡ Thus for small values of the drain voltage V_D the channel will again act as a resistor, but its resistance will be larger because the cross-sectional area available for current flow will have decreased due to the increased width W of the depletion regions. This is evident from the experimental data corresponding to $V_G = -1$ v, -2 v, etc. in Figure 8.3.

As V_D is increased, the resistance of the channel will increase. When V_D reaches a large enough value, the depletion regions will again touch near the drain region. This will take place when $V_D = V_{D\text{sat}}$, where

$$V_{D\text{sat}} = \frac{qN_D d^2}{8K_s \epsilon_0} - \phi_B + V_G \tag{8.4}$$

which is the voltage required for the onset of saturation in the presence of a gate voltage. It is evident that the application of a gate voltage V_G *lowers the drain voltage required for the onset of saturation by an equal amount*.§ (Recall that V_G is negative for an *n*-channel device.)

As the drain voltage is further increased beyond $V_{D\text{sat}}$, the point X where the depletion regions touch will again merely move slightly toward the source but the voltage drop from the source to the point X will remain the same, $V_{D\text{sat}}$. Accordingly, the drain current will also remain at its value corresponding to the onset of saturation $I_{D\text{sat}}$. This value, however, will be lower than in the $V_G = 0$ case because the voltage drop from source to point X, $V_{D\text{sat}}$, itself is lower. This is evident from the experimental data in Figure 8.3.

Thus we can distinguish two different regions of the current-voltage relationship of field-effect transistors. In one region, when V_D is small, the cross-sectional area of the channel is practically independent of V_D

---

† This assumes that the movement of the point X toward the source is negligible. The validity of this assumption will be discussed in a later section.
‡ Usually the same bias is applied to both gates.
§ The above as well as other formulas are summarized at the end of this chapter in Table 8.1 for both *n*- and *p*-channel devices.

and the current-voltage characteristics are essentially ohmic or linear. We refer to this limiting region of operation of field-effect transistors as the *linear region*. In the other extreme, for $V_D \geqslant V_{D\text{sat}}$, the current saturates at $I_{D\text{sat}}$. We will refer to this region of operation of field-effect transistors as the *saturation region*.

8.2 CHARACTERISTICS OF JUNCTION FIELD-EFFECT TRANSISTORS

a. Current-Voltage Relationship[1]

Let us now consider a junction field-effect transistor before the onset of saturation as shown in Figure 8.4. The voltage drop across the elemental section of the channel is given by

$$dV = I_D\, dR = \frac{I_D\, dy}{q\mu_n N_D Z[d - 2W(y)]} \qquad (8.5)$$

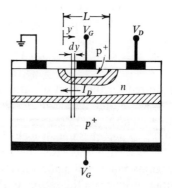

Fig. 8.4 The elemental section of the channel employed in the derivation of the current-voltage characteristics of junction field-effect transistors.

in analogy to Equation 8.1 except that L is replaced by dy. The depletion region width at distance y from the source is given by

$$W(y) = \sqrt{\frac{2K_s\epsilon_0[V(y) + \phi_B - V_G]}{qN_D}}. \qquad (8.6)$$

Substituting Equation 8.6 into 8.5 and integrating between the source (where $y = 0$ and $V = 0$†) and the drain (where $y = L$ and $V = V_D$) leads

† We define the potential at the source as our ground potential. This analysis neglects the series resistance between source contact and the beginning of the channel ($y = 0$), and between drain contact and the end of the channel ($y = L$).

Characteristics of Junction Field-Effect Transistors

to the fundamental equation of field-effect transistors:

$$I_D = G_o \left\{ V_D - \frac{2}{3}\sqrt{\frac{8K_s\epsilon_0}{qN_Dd^2}} \left[(V_D + \phi_B - V_G)^{3/2} - (\phi_B - V_G)^{3/2}\right] \right\} \quad (8.7)$$

where

$$G_o = \frac{Zq\mu_n N_D d}{L} \quad (8.8)$$

is the conductance of the metallurgical channel, i.e., the conductance of the n-type layer between the two p-type regions discounting the presence

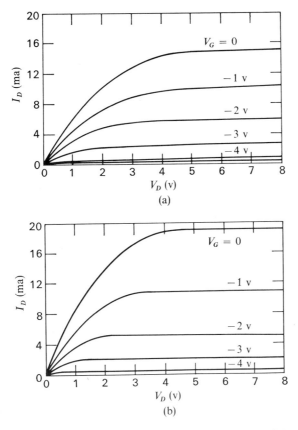

Fig. 8.5 Current-voltage characteristics of the n-channel junction field-effect transistor. (a) Experiment (same as Fig. 8.3). (b) Theory—Eq. 8.7 (no series resistance).

of the depletion regions altogether. In Figure 8.5a we show the experimentally measured current-voltage characteristics of the field-effect transistor used in this chapter as an example. Below it, in Figure 8.5b, we show the family of characteristics calculated from Equation 8.7, using the known values of structural parameters. Equation 8.7 is valid only below saturation. Thus the curves shown were calculated for $0 \leqslant V_D \leqslant V_{D\text{sat}}$ using Equation 8.7; beyond $V_{D\text{sat}}$ the current was taken to be constant, in line with our earlier argument.

Of particular importance is the form of the above equation in the linear and in the saturation regions. For small drain voltages, i.e., for $V_D \ll \phi_B - V_G$, the bracketed terms can be expanded leading to the simple formula,

$$I_D \cong G_o \left[1 - \sqrt{\frac{8K_s\epsilon_0(\phi_B - V_G)}{qN_Dd^2}} \right] V_D \qquad \text{[Linear region]}. \qquad (8.9)$$

It is evident that this expression gives the current-voltage relationship of a resistor whose resistance increases with gate voltage.

The channel conductance g is given by

$$g = \frac{\partial I_D}{\partial V_D}\bigg|_{V_G=\text{const}}. \qquad (8.10)$$

In the linear region, g is given from Equation 8.9 as

$$g = G_o \left[1 - \sqrt{\frac{8K_s\epsilon_0(\phi_B - V_G)}{qN_Dd^2}} \right] \qquad \text{[Linear region]}. \qquad (8.11)$$

In Figure 8.6, we show the experimentally measured channel conductance in the linear region of our device in comparison with calculations based on Equation 8.11 for this device. (Also shown are curves which were corrected for the presence of series resistances between source and drain contacts and the channel region. This correction will be discussed later.) As the gate bias is increased, the conductance decreases until finally, at a certain value of gate voltage, the conductance vanishes altogether. This voltage is called the *turn-off voltage* and it corresponds to the reverse bias that must be applied to the two gate junctions in order to deplete all of the channel region. Accordingly, the turn-off voltage can be calculated from the condition $W = d/2$ to be

$$V_T = -\frac{qN_Dd^2}{8K_s\epsilon_0} + \phi_B. \qquad (8.12)$$

Characteristics of Junction Field-Effect Transistors

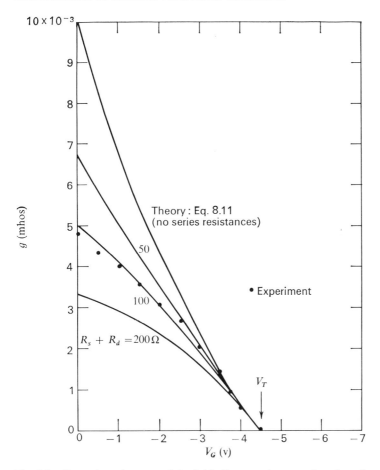

Fig. 8.6 Channel conductance of the field-effect transistor as a function of gate voltage in the linear region (V_D small).

The drain current in the saturation region $I_{D\text{sat}}$ can be calculated from Equation 8.7 by evaluating the drain current at the limit of the applicability of this equation, i.e., at $V_D = V_{D\text{sat}}$. Substitution of Equation 8.4 into 8.7 yields the saturation current,

$$I_{D\text{sat}} = G_o \left\{ \left[\frac{2}{3} \sqrt{\frac{8K_s \epsilon_0 (\phi_B - V_G)}{qN_D d^2}} - 1 \right] (\phi_B - V_G) \right. $$
$$\left. + \frac{1}{3} \frac{qN_D d^2}{8K_s \epsilon_0} \right\} \quad \text{[Saturation region]}. \tag{8.13}$$

b. Transconductance

An important property of field-effect transistors is the transconductance defined by

$$g_m \equiv \left.\frac{\partial I_D}{\partial V_G}\right|_{V_D=\text{const}}. \tag{8.14}$$

The transconductance represents the change of drain current at a given drain voltage upon a change in gate voltage. It can be readily evaluated by differentiating Equation 8.7. This yields

$$g_m = G_0\sqrt{\frac{8K_s\epsilon_0}{qN_Dd^2}}\,[\sqrt{V_D + \phi_B - V_G} - \sqrt{\phi_B - V_G}]. \tag{8.15}$$

The transconductance in the linear region can be obtained by expanding the bracketed terms. This yields

$$g_m = G_0\sqrt{\frac{8K_s\epsilon_0}{qN_Dd^2}}\frac{V_D}{2\sqrt{\phi_B - V_G}} \quad [\text{Linear region}]. \tag{8.16}$$

Conversely, the transconductance in the saturation region can be calculated by inserting $V_D = V_{D\text{sat}}$ into Equation 8.15. This yields

$$g_{m\text{sat}} = G_0\left[1 - \sqrt{\frac{8K_s\epsilon_0(\phi_B - V_G)}{qN_Dd^2}}\right] \quad [\text{Saturation region}]. \tag{8.17}$$

Comparison with Equation 8.11 shows that *the transconductance in the saturation region exactly equals the conductance in the linear region.* This important result is given experimental verification by a comparison of the transconductance data shown in Figure 8.7 with the conductance measurements shown in Figure 8.6.

c. Gate Leakage Current

Because the gate is reverse biased with respect to the channel, the current flowing to the gate terminal is very small; it is the reverse-bias leakage current of a *p-n* junction. Thus the gate impedance, i.e., the input impedance of a field-effect transistor, is very high. This high input impedance is one of the very important distinguishing characteristics of field-effect transistors as compared to junction transistors.

The ordinary room-temperature leakage currents of reverse-biased silicon *p-n* junctions prior to breakdown are of the order of picoamps to nanoamps. However, surface effects can lead to drastic increases in

Modifications of the Simple Theory

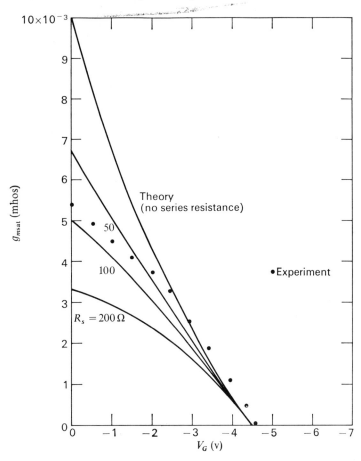

Fig. 8.7 Transconductance of the field-effect transistor as a function of gate voltage in the saturation region. ($V_D > V_{Dsat}$).

junction leakage currents as discussed in Chapter 10. In the same manner, they can lead to similar increases in the gate leakage current of junction field-effect transistors. Thus surface effects can degrade the high input impedance of field-effect transistors.

8.3 MODIFICATIONS OF THE SIMPLE THEORY

a. Graded Channel Regions

In Section 8.2 we considered a field-effect transistor in which the impurity concentration in the channel region was uniform; and we employed the one-sided step-junction approximation. This is a reasonably good approximation for field-effect transistors fabricated in the manner described in the beginning of the chapter since the junctions are usually shallow. However, if the diffusions are deeper and the channel is located further away from the surface, we may approach the range of validity of the linearly graded junction approximation.

Also, field-effect transistors can be made entirely by diffusion techniques, without recourse to epitaxial growth. Thus, for instance, they can be made by successive diffusions of impurities of opposite type much the same way as the emitter and base regions of a double-diffused junction transistor are formed. In such a case the impurity distribution in the channel region is similar to the impurity distribution in the base of a transistor and is extremely graded.

However, it has been shown theoretically[3] that the general characteristics of the junction field-effect transistor are quite independent of the particular impurity distribution. Thus the simplest analysis, that for the uniform impurity distribution presented in the previous section, provides most of the general features of the analysis for an arbitrary distribution to a fairly good approximation.

b. Cut-Off Frequency for Transconductance

When the gate voltage is changed by an amount ΔV_G, the width of the depletion regions changes. As a result, the drain current will also change. Some of the additional drain current will be used to make up the change in charge contained within the depletion regions of the gate junctions. The response time of the field-effect transistor t_o can then be defined as that time in which the change in the drain current makes up the change in the total charge on the gate, that is,

$$t_o \Delta I_D = \Delta Q_G A_G. \tag{8.18}$$

Thus it follows that this response time is given by

$$t_o = \frac{\Delta Q_G A_G}{\Delta I_D} = \frac{\Delta Q_G A_G}{\Delta V_G} \frac{\Delta V_G}{\Delta I_D} = \frac{C_G}{g_m}. \tag{8.19}$$

Here C_G is the *total* gate capacitance of the device, given by

$$C_G = 2ZL \frac{K_s \epsilon_0}{\overline{W}} \tag{8.20}$$

where $\overline{W}$ is the average depletion region width. The maximum frequency

Modifications of the Simple Theory

of operation of the field-effect transistor is then given by the frequency corresponding to this charging time constant,

$$f_o = \frac{1}{t_o} = \frac{g_m}{C_G}.\dagger \qquad (8.21)$$

A simple estimate of the upper limit of f_o can be obtained from the ratio of the maximum transconductance to the minimum gate capacitance. The maximum transconductance, from the previous section, is G_o. The minimum gate capacitance is obtained when the depletion region reaches its maximum width, which is one-half of the channel depth d. Thus,

$$f_o < \frac{q\mu_n N_D d^2}{4K_s \epsilon_0 L^2}. \qquad (8.22)$$

Note that this limiting frequency is proportional to the mobility. Because the mobility of electrons in silicon is approximately twice the mobility of holes, an n-channel device promises twice as high a frequency range as a p-channel device for the same geometric factors and doping levels.

c. Source-to-Drain Resistance in Saturation

We have already seen in the previous section that in the saturation region the potential at the end of the channel, at the point X of Figure 8.2, will be fixed at precisely the value of $V_{D\text{sat}}$ corresponding to the applied gate voltage. This is so because the point X itself is where the two depletion regions just touch. Hence, the reverse bias across the gate junctions at this point is fixed by the condition that $W = d/2$ there.

As the drain voltage is increased further, the reverse bias between gate and the drain region itself is also increased; hence, the width of the depletion region near the drain will also increase. As a result, the point X will move toward the source as indicated in Figure 8.2c. The voltage at the point X remains at the same value but the length L from the source to the point X shortens; thus it is evident that the drain current will increase at a given gate voltage as the drain voltage is increased. This results in an upward tilting of the current-voltage characteristics beyond saturation which is particularly prominent for devices with small channel length L.

This phenomenon is quite analogous to the Early effect discussed in connection with junction transistors. In both cases, the increase in current takes place because the current path is shortened by the widening of a reverse-biased depletion region.

† More rigorous considerations yield $1/2\pi t_o$ for this frequency limitation.

d. Effect of Series Resistance

In the above calculations we considered only the resistance of that portion of the channel which can be modulated by the application of a reverse bias to the gate. In reality, there are series resistances present, both near the source and near the drain, which interpose an *IR* drop between the source and drain contacts and the channel. These series resistances are illustrated schematically in Figure 8.8.

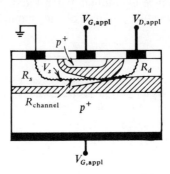

Fig. 8.8 Illustration of the series resistances due to the unmodulated portions of the channel near source and drain.

The effect of these series resistances on the channel conductance in the linear region can be readily calculated by noting that

$$\frac{1}{g(\text{obs})} = \frac{1}{g} + R_s + R_d \tag{8.23}$$

where g is the true channel conductance while $g(\text{obs})$ is the conductance observed experimentally. R_s and R_d are the series resistances near the source and the drain, respectively, as indicated in Figure 8.8. Thus,

$$g(\text{obs}) = \frac{g}{1 + (R_s + R_d)g} \tag{8.24}$$

which shows that the observed conductance will be reduced due to the two series resistances. This reduction was taken into account in Figure 8.6 for various series resistance values. The value $(R_s + R_d) \simeq 100 \ \Omega$, which provides a good fit with the experimental data, is in fact a very reasonable one for this particular device geometry.

Let us now consider the effect of the series resistance near the source region, R_s, on the transconductance in the saturation region. Because of this resistance, the potential at the beginning of the channel will not be zero, as assumed in the previous treatment, but will have some finite value V_s. Thus the *effective* gate voltage will be

$$V_G = V_{G,\text{appl}} - V_s. \qquad (8.25)$$

As a result, the observed transconductance is given by

$$g_m(\text{obs}) = \frac{dI_D}{dV_{G,\text{appl}}} = \frac{dI_D}{d[V_G + V_s]} \qquad (8.26)$$

which, in turn, yields

$$g_m(\text{obs}) = \frac{1}{\dfrac{dV_G}{dI_D} + \dfrac{dV_s}{dI_D}} \qquad (8.27)$$

and hence,

$$g_m(\text{obs}) = \frac{g_m}{1 + R_s g_m}. \qquad (8.28)$$

This last equation shows that the observed transconductance in the saturation region will be reduced due to the presence of a series resistance near the source from that attainable in the absence of such a series resistance. This reduction is shown in Figure 8.7 for various values of R_s.

The series resistance near the drain will act in a different manner. Because of the *IR* drop across this resistance, the drain voltage required to bring about saturation of the drain current will be larger than without it. However, since beyond that voltage, i.e., for $V_D > V_{D\text{sat}}$, the magnitude of V_D has no significant effect on the drain current, the drain series resistance will have no further effect either.

READING REFERENCES

J. T. Wallmark, "The Field-Effect Transistor—A Review," *RCA Rev.*, **24**, 641 (1963).

L. J. Sevin, *Field-Effect Transistors*, McGraw-Hill Book Co., 1965.

REFERENCES CITED

1. W. Shockley, "A Unipolar 'Field-Effect' Transistor," *Proc. IRE*, **40**, 1365 (1952).
2. G. C. Dacey and I. M. Ross, "Unipolar 'Field-Effect' Transistor," *Proc. IRE*, **41**, 970 (1953); G. C. Dacey and I. M. Ross, "The Field-Effect Transistor," *Bell System Tech. J.*, **34**, 1149 (1955).
3. R. S. C. Cobbold and F. N. Trofimenkoff, "Theory and Application of the Field-Effect Transistor," *Proc. IEE*, **111**, 1981 (1964).

PROBLEMS

8.1 Prepare a table comparing the electrical characteristics of junction transistors and junction field-effect transistors.

8.2 Rearrange the current-voltage relationship, and the expressions for conductance and transconductance by employing the expression for the turn-off voltage V_T, Equation 8.12.

8.3 Rederive the current-voltage relationship, Equation 8.7, in the presence of a constant (unmodulated) series resistance near the source, R_s, and near the drain, R_d. Discuss the effect of each. In particular, show that V_{Dsat} is independent of R_s.

8.4 Derive an expression for the drain conductance $g = dI_D/dV_D$ at a given V_G, in the saturation region. Assume that this conductance is due to the thickening of the depletion region near the drain, and approximate the latter by the one-dimensional one-sided step-junction formula.

8.5 Derive an expression giving the electric field along the channel, and the carrier velocity in the channel. Examine the validity of the use of a carrier velocity which is proportional to the electric field. Discuss qualitatively how the current-voltage characteristics may be affected by the real carrier velocity versus electric field relationship.

8.6 Derive expressions giving the variation of the conductance in the linear region and of the transconductance in the saturation region with temperature, at a fixed gate voltage.

8.7 It is often stated that junction field-effect transistors are insensitive to the effect of irradiation because their operation is independent of the lifetime. Recalling the "carrier-removal" effect (Chapter 5), derive an expression giving the channel conductance in the linear region as a function of radiation dose, for low radiation doses. Using the data in Figure 5.17, estimate the electron dose at which the channel conductance of the field-effect transistor used in the present chapter for illustration is reduced by 10%.

TABLE 8.1
IMPORTANT FORMULAS FOR JUNCTION FIELD-EFFECT TRANSISTORS

	n-channel $V_D > 0$, $V_G < 0$; I_D flows from drain to source	p-channel $V_D < 0$, $V_G > 0$; I_D flows from source to drain
Current-voltage characteristics	$I_D = G_o \left\{ V_D - \dfrac{2}{3} \sqrt{\dfrac{8K_s\epsilon_0}{qN_Dd^2}} \times [(V_D + \phi_B - V_G)^{3/2} - (\phi_B - V_G)^{3/2}] \right\}$ where $G_o = \dfrac{Z}{L} q\mu_n N_D d$	$I_D = G_o \left\{ -V_D - \dfrac{2}{3} \sqrt{\dfrac{8K_s\epsilon_0}{qN_Ad^2}} \times [(V_G + \phi_B - V_D)^{3/2} - (V_G + \phi_B)^{3/2}] \right\}$ where $G_o = \dfrac{Z}{L} q\mu_p N_A d$
Saturation voltage	$V_{D\mathrm{sat}} = \dfrac{qN_Dd^2}{8K_s\epsilon_0} - \phi_B + V_G$	$V_{D\mathrm{sat}} = -\dfrac{qN_Ad^2}{8K_s\epsilon_0} + \phi_B + V_G$
Turn-off voltage	$V_T = -\dfrac{qN_Dd^2}{8K_s\epsilon_0} + \phi_B$	$V_T = \dfrac{qN_Ad^2}{8K_s\epsilon_0} - \phi_B$
Conductance (linear region) Transconductance (saturation)	$\begin{Bmatrix} g_{\mathrm{linear}} \\ g_{m\mathrm{sat}} \end{Bmatrix} = G_o \left[1 - \sqrt{\dfrac{8K_s\epsilon_0(\phi_B - V_G)}{qN_Dd^2}} \right]$	$\begin{Bmatrix} g_{\mathrm{linear}} \\ g_{m\mathrm{sat}} \end{Bmatrix} = G_o \left[1 - \sqrt{\dfrac{8K_s\epsilon_0(\phi_B + V_G)}{qN_Ad^2}} \right]$
Saturation current	$I_{D\mathrm{sat}} = G_o \left\{ \left[\dfrac{2}{3} \sqrt{\dfrac{8K_s\epsilon_0(\phi_B - V_G)}{qN_Dd^2}} - 1 \right] \times (\phi_B - V_G) + \dfrac{1}{3}\dfrac{qN_Dd^2}{8K_s\epsilon_0} \right\}$	$I_{D\mathrm{sat}} = G_o \left\{ \left[\dfrac{2}{3} \sqrt{\dfrac{8K_s\epsilon_0(\phi_B + V_G)}{qN_Ad^2}} - 1 \right] \times (\phi_B + V_G) + \dfrac{1}{3}\dfrac{qN_Ad^2}{8K_s\epsilon_0} \right\}$
Maximum frequency	$f_o = \dfrac{g_m}{C_G} \leq \dfrac{q\mu_n N_D d^2}{2K_s\epsilon_0 L^2}$	$f_o = \dfrac{g_m}{C_G} \leq \dfrac{q\mu_p N_A d^2}{2K_s\epsilon_0 L^2}$
Effect of series resistances	Linear region $g(\mathrm{obs}) = \dfrac{g}{1 + (R_s + R_d)g}$ Saturation region $g_{m\mathrm{sat}}(\mathrm{obs}) = \dfrac{g_{m\mathrm{sat}}}{1 + R_s g_{m\mathrm{sat}}}$	

PART III

SURFACE EFFECTS AND SURFACE-CONTROLLED DEVICES

- **Theory of Semiconductor Surfaces**
- **Surface Effects on *p-n* Junctions**
- **Surface Field-Effect Transistors**
- **Properties of the Si-SiO$_2$ System**

- CHARACTERISTICS OF SURFACE SPACE-CHARGE REGIONS—EQUILIBRIUM CASE
- THE IDEAL MIS STRUCTURE
- WORK FUNCTION DIFFERENCE; CHARGES, AND STATES

9
Theory of Semiconductor Surfaces

In Part II we have considered semiconductor devices with the implicit assumption that their characteristics are determined strictly by bulk phenomena. In reality, surface effects often completely dominate the characteristics of semiconductor devices. In fact, as we had mentioned in the Introduction, one of the principal reasons for the success of the planar technology is that planar junctions are covered by a thermally grown silicon dioxide layer. This reduces—but by no means eliminates—many surface effects and therefore results in better control of device characteristics.

The properties of the thermally oxidized silicon interface have been studied very extensively; perhaps more extensively than any other system in the long history of semiconductor surface research. Because of this, and because of their great importance in semiconductor device technology, we will now consider surface effects in some detail.

Surface effects on *p-n* junctions are primarily due to the fact that ionic charges outside the semiconductor surface will induce an image charge in the semiconductor and thereby lead to the formation of *surface space-charge regions*. This is illustrated schematically in Figure 9.1 where we show the idealized plane *p-n* junction structure that was the basis of our previous studies. If a surface space-charge region is formed, it will modify the junction space-charge region and can lead to changes in junction characteristics.

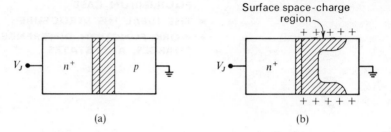

Fig. 9.1 Illustration of the role of surface effects in semiconductor devices.
(a) Idealized plane p-n junction.
(b) Same, with surface space charge induced by positive ions.

In this chapter we will first study the characteristics of surface space-charge regions in thermal equilibrium. After that we will study the metal-insulator-semiconductor (MIS) structure,† which has proved to be extremely useful in the study of semiconductor surfaces.[1] We will first consider its characteristics in the ideal case; then we will extend our consideration to include the effect of metal-semiconductor work-function differences, interface and oxide charges, and interface states.

9.1 CHARACTERISTICS OF SURFACE SPACE-CHARGE REGIONS—EQUILIBRIUM CASE[2]

The three experimental structures customarily employed in the study of surface effects and of the characteristics of surface space-charge regions are illustrated in Figure 9.2. These structures are (a) the metal-insulator-semiconductor capacitor structure; (b) the gate or field-plate controlled p-n junction; and (c) the metal-insulator-semiconductor surface field-effect transistor. In all three cases we illustrate the case of a p-type substrate, with a large positive voltage applied to the field plate.

An important difference exists between the first structure and the second two. In the case of a metal-insulator-semiconductor capacitor, no d-c current flow is possible across the space-charge region. This is because any such current flow would have to go through the insulator. In contrast, current flow across the surface space-charge region is possible in structures (b) and (c) because in these cases a contact is made to the surface space-charge region through the n^+ regions. Thus, the surface space-charge

---

† Because in most of the experimental studies the insulator has been silicon dioxide, the term 'metal-oxide-semiconductor (MOS) structure' will be used interchangeably with 'metal-insulator-semiconductor (MIS) structure'.

Characteristics of Surface Space-Charge Regions

region in the case of the metal-insulator-semiconductor capacitor will be in *thermal equilibrium;* i.e., $np = n_i^2$, and the Fermi level will be constant throughout the surface space-charge region.† In this chapter we will restrict our attention to this case. In order to treat characteristics of surface space-charge regions associated with *p-n* junctions under bias, such as in case (b), we will have to extend our consideration to *nonequilibrium* conditions. This will be done in the next chapter.

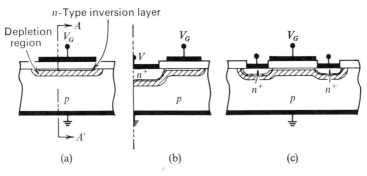

Fig. 9.2 The experimental structures used to study surface effects.
(a) The metal-insulator-semiconductor capacitor structure.
(b) The gate-controlled *p-n* junction.
(c) The metal-insulator-semiconductor surface field-effect transistor.
   In all cases $V_G \gg 0$.

The energy band diagram of a metal-oxide-semiconductor structure with a *p*-type semiconductor is shown in Figure 9.3 for three bias conditions. Regardless of the value of the gate voltage, the Fermi level in the semiconductor remains constant since equilibrium holds. In Figure 9.3a, we show the conditions corresponding to a negative voltage applied to the metal gate. This negative potential will attract a positive charge in the semiconductor which, in the case of a *p*-type semiconductor, will consist of an enhanced concentration—*accumulation*—of holes near the oxide-silicon interface. The corresponding charge distribution is shown in the lower half of the figure.‡

If a small positive voltage is applied to the gate, a negative charge will

---

† It should be noted that deviations from the equilibrium condition are possible also in the case of the simple MIS structure. Such deviations can take place if the measurements are made under transient conditions, or if the insulator is imperfect and current can leak across it. Although both of these conditions can be observed experimentally,[3] we will not consider them in this discussion.

‡ The conduction band of the silicon-dioxide layer is also indicated in this figure. This conduction band is discussed in greater detail in Section 9.3.

be induced in the semiconductor (Figure 9.3b). This, at first, will be due to holes being pushed away from the vicinity of the interface, leaving behind a *depletion region* consisting of uncompensated acceptor ions as shown in the lower half of the figure. The charge per unit area contained in the semiconductor Q_s will then be given by the charge contained within this depletion region,

$$Q_s = -qN_A x_d, \qquad (9.1)$$

where x_d is the width of the surface depletion region.

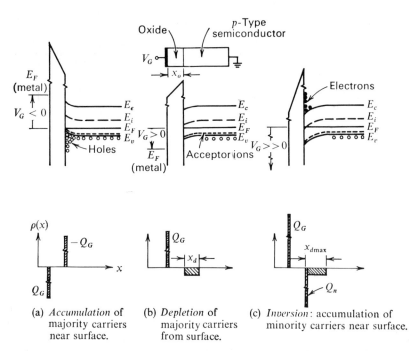

(a) *Accumulation* of majority carriers near surface.

(b) *Depletion* of majority carriers from surface.

(c) *Inversion*: accumulation of minority carriers near surface.

Fig. 9.3 Energy bands and charge distribution in an MOS structure under various bias conditions, in the absence of surface states and work function difference.[3]

If we increase the positive potential applied to the gate, the width of the surface depletion region will first increase. Correspondingly, the total electrostatic potential variation in the silicon, as represented by the bending of the energy bands, will also increase as shown in Figure 9.3c. However, as the bands are bent further, the conduction band will eventually come close to the Fermi level. When this happens, the concentration of electrons near the interface will suddenly increase very sharply. After this, most of the additional negative charge induced in the semiconductor will consist

Characteristics of Surface Space-Charge Regions

of the charge Q_n due to the electrons in a very narrow n-type *inversion layer*.†

Once an inversion layer is formed, the width of the surface depletion region reaches a maximum. This is because once the bands are pulled down far enough for strong inversion to occur, even a very small increase in band bending, corresponding to a very small increase in depletion region width, will result in a very large increase in the charge contained within the inversion layer. Thus, under such *strong inversion* conditions the charge per unit area induced in the semiconductor will be given by

$$Q_s = Q_n - qN_A x_{d\max} \tag{9.2}$$

where $x_{d\max}$ denotes the *maximum width of the surface depletion region*.

For most purposes, we are interested only in the cases of depletion and inversion. These cases can be described to an excellent approximation by the *depletion approximation* that we have already employed in our study of p-n junctions.

When the semiconductor is depleted and the charge within the semiconductor is given by Equation 9.1, integration of Poisson's equation yields the distribution of the electrostatic potential in the surface depletion region as

$$\phi = \phi_s \left(1 - \frac{x}{x_d}\right)^2 \tag{9.3}$$

where the *surface potential* ϕ_s, which designates the total bending of the energy bands from the bulk of the semiconductor to the surface, is given by

$$\phi_s = \frac{qN_A x_d^2}{2K_s \epsilon_0}. \tag{9.4}$$

We should note that this potential distribution is identical with the one we derived in Chapter 6 for the case of a one-sided step junction. Indeed, we will see later that the theory of surface space-charge regions for the depletion and inversion cases is analogous to the theory of one-sided step junctions in almost every detail.

The distribution given by Equation 9.3 is shown in Figure 9.4 in comparison to exact numerical calculations[4] of the potential distribution in the surface space-charge region for a strongly inverted surface. The agreement is clearly reasonable, except very near the surface. (All calculations in this chapter are for silicon at 300°K, unless otherwise specified.)

† The inversion layer is narrow because the minority carrier concentration drops to $\sim 10\%$ of its value at the interface over a distance $\sim 3kT/q\mathscr{E}_s$ where $\mathscr{E}_s$ is the electric field in the semiconductor, at the interface. This distance, 10 to 100 Å, is much smaller than the width of the surface depletion region.

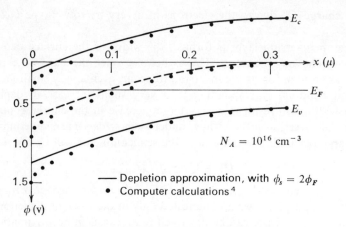

Fig. 9.4 Potential distribution in the surface depletion region.

One particularly interesting question concerns the point after which the charge due to electrons in the inversion layer becomes significant. A good approximate criterion for the onset of this condition of *strong inversion*† is that the electron *concentration* (per unit volume) near the surface should exceed the concentration of the substrate impurity ions, i.e.,

$$n_s = N_A \tag{9.5}$$

for the case of a *p*-type semiconductor. Under this condition, the Fermi level at the surface will be above the intrinsic Fermi level by as much as it is below the intrinsic Fermi level in the bulk. Accordingly, the total band-bending at the onset of strong inversion will be

$$\phi_s(\text{inv}) = 2\phi_F. \tag{9.6}$$

As we discussed earlier, the surface depletion region width reaches a maximum when the surface becomes strongly inverted. Accordingly, the maximum width of the surface depletion region can be estimated using the depletion approximation, corresponding to the value of the surface potential at the onset of strong inversion, $\phi_s(\text{inv})$ as defined above. This yields

$$x_{d\max} = \sqrt{\frac{2K_s\epsilon_0\phi_s(\text{inv})}{qN_A}}. \tag{9.7}$$

Note the similarity between this expression and that giving the width

† Strictly speaking, the surface becomes inverted when the minority carrier concentration at the surface equals the majority carrier concentration at the surface, i.e., $n_s = p_s = n_i$. However, this is not a useful criterion because the corresponding charge of minority carriers per unit area is immeasurably small.

of the zero-bias depletion region of a one-sided step junction (Chapter 6). The two expressions are similar with $\phi_s(\text{inv}) = 2\phi_F$ taking the place of the built-in voltage ϕ_B. This analogy is indeed reasonable: when the surface is inverted, a very thin n-type layer is formed, separated from the p-type substrate by a depletion region. The one difference between the n^+p step-junction case considered earlier and the inversion layer-substrate junction that we are dealing with here is that, in the former, the n-type conductivity was brought about by a metallurgical process; i.e., the electrons were

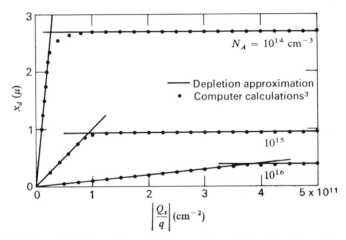

Fig. 9.5 Surface depletion region width versus charge induced in the semiconductor.

introduced into the semiconductor by the introduction of donor ions. In contrast, in the present case the n-type layer is induced by the electric field applied with the field plate. Thus this junction, rather than being a metallurgical junction, is a *field-induced junction*.

We can readily calculate the depletion region width x_d as a function of the charge per unit area induced in the silicon Q_s. Such calculations based both on the depletion approximation and on more exact numerical computations[3] are shown in Figure 9.5. Note that, in accordance with our earlier qualitative argument, the depletion region width first increases as an increasing amount of charge is induced in the silicon. The rate of increase—the slope of the straight lines—is of course dependent on the impurity concentration. Eventually, when the surface is inverted and a field-induced junction is formed, the surface depletion region reaches a maximum width. This maximum surface depletion region width $x_{d\max}$ is only a function of the impurity concentration, the same way as the zero-bias depletion region width of a one-sided step junction is only a function of the substrate impurity concentration. The relationship between $x_{d\max}$

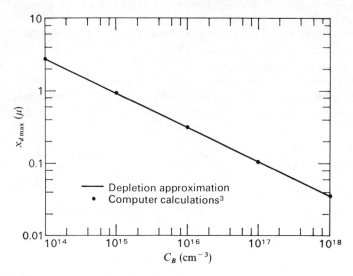

Fig. 9.6 The maximum width of the surface depletion region as a function of substrate impurity concentration.

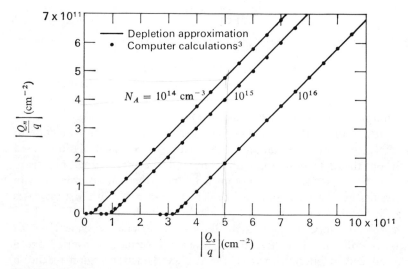

Fig. 9.7 The magnitude of the charge of the minority carriers within the inversion layer as a function of the magnitude of the total charge induced within the semiconductor.

and impurity concentration, based on both the depletion approximation and numerical computations, is shown in Figure 9.6.

An important quantity is the charge per unit area contained within the surface depletion region at and after the onset of strong inversion. This is given by

$$Q_B \equiv -qN_A x_{d\max} \tag{9.8}$$

for a *p*-type semiconductor. According to Equation 9.2, the total charge in the silicon after strong inversion will then be given by

$$Q_s = Q_n + Q_B \tag{9.9}$$

or

$$Q_n = Q_s - Q_B. \tag{9.10}$$

This relationship is shown in Figure 9.7 for three different doping concentrations. Again, the depletion approximation is compared with exact numerical computations. The agreement in this and in the previous figures demonstrates that the depletion approximation provides an excellent description of the surface space-charge region in the depletion and inversion regimes.

9.2 THE IDEAL MIS (OR MOS) STRUCTURE

a. Capacitance-Voltage Characteristics

In the absence of any contact potential or work-function differences between metal and semiconductor (we will discuss these in Section 9.3), any applied gate voltage will partly appear across the oxide and partly across the silicon. Thus,

$$V_G = V_o + \phi_s \tag{9.11}$$

where V_o and ϕ_s denote the potential variation across oxide and semiconductor, respectively.

In the absence of any charges located at the interface between the oxide and the semiconductor, Gauss' law requires that the electric displacement should be continuous at this interface or

$$K_o \mathscr{E}_o = K_s \mathscr{E}_s \text{ at the interface.} \tag{9.12}$$

If there are no charges present within the oxide, the electric field in it is uniform and is given by

$$\mathscr{E}_o = \frac{V_o}{x_o}, \tag{9.13}$$

where x_o is the oxide thickness. The electric field at the silicon surface, from Gauss' law, equals

$$\mathcal{E}_s = -\frac{Q_s}{K_s \epsilon_0}. \tag{9.14}$$

A combination of these three equations yields the voltage drop across the oxide as

$$V_o = -\frac{x_o}{K_o \epsilon_0} Q_s = -\frac{Q_s}{C_o} \tag{9.15}$$

where $C_o \equiv K_o \epsilon_0 / x_o$ is the capacitance per unit area of the oxide layer.

The gate voltage then will be related to the characteristics of the surface space-charge region by

$$V_G = -\frac{Q_s}{C_o} + \phi_s. \tag{9.16}$$

The simplest measurable electrical characteristic of an MOS structure is its small-signal capacitance. If Q_G is the charge per unit area on the gate, this capacitance is given by

$$C \equiv \frac{dQ_G}{dV_G} = -\frac{dQ_s}{dV_G} = -\frac{dQ_s}{-\frac{dQ_s}{C_o} + d\phi_s} \tag{9.17}$$

or

$$C = \frac{1}{\frac{1}{C_o} + \frac{1}{C_s}} \tag{9.18}$$

where

$$C_s \equiv -\frac{dQ_s}{d\phi_s} = \frac{K_s \epsilon_0}{x_d} \tag{9.19}$$

is the capacitance per unit area of the surface space-charge region in the semiconductor. Thus, the capacitance of the MOS structure is given by the series combination of C_o and C_s. By eliminating x_d, we obtain the formula for the capacitance of an MOS structure:

$$\frac{C}{C_o} = \frac{1}{\sqrt{1 + \frac{2K_o^2 \epsilon_0}{q N_A K_s x_o^2} V_G}} \tag{9.20}$$

which predicts that the capacitance will fall with the square root of the gate voltage *while the surface is being depleted.*

The Ideal MIS (or MOS) Structure

When the gate voltage is zero or negative, no depletion region exists. Thus the depletion approximation, and with it the above formula, lose their meaning. However, the capacitance can be readily obtained by considering that when the surface is accumulated, the semiconductor merely acts as a resistor in series with the oxide capacitance. Thus the measured capacitance will be simply C_o.

In the other extreme, when strong inversion sets in, the width of the depletion region will not increase with further increase in gate voltage.

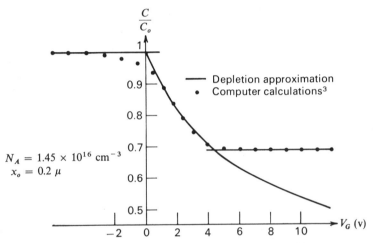

Fig. 9.8 The capacitance-voltage characteristics of an MOS structure.

This will take place at a gate voltage corresponding to a charge per unit area Q_B induced in the silicon and a surface potential $\phi_s(\text{inv}) = 2\phi_F$. Substitution of these quantities into Equation 9.16 yields the gate voltage at the onset of strong inversion, referred to as the *turn-on voltage* for reasons that will become clear later,

$Q_B = -q N_A x_{d\,max}$

$$V_T \equiv -\frac{Q_B}{C_o} + \phi_s(\text{inv}). \tag{9.21}$$

Thus the capacitance will level off and remain constant at a value given by Equation 9.20 for $V_G = V_T$. The capacitance-voltage characteristic of a particular metal-silicon dioxide-silicon structure is illustrated in Figure 9.8 based on both the depletion approximation and exact computer calculations. The depletion approximation evidently shows all the correct features of the more exact theory.

It should be pointed out that while we have used a *p*-type substrate in the preceding discussions, all of the above considerations are equally valid to the case of an n-type substrate with proper changes in symbols (e.g., Q_n to Q_p) and signs. In particular, the capacitance-voltage characteristics will have identical shapes but will be mirror images of each other.

Table 9.1 at the end of this chapter summarizes the important formulas involved in the analysis of surface space-charge regions for both p- and n-type semiconductors.

b. Frequency Effects

In the preceding calculation of the capacitance we had assumed that upon a change in gate voltage all of the incremental charge appears at the edge of the depletion region. This consideration led to the simple formula for the capacitance of the semiconductor space-charge region,

$$C_s \equiv -\frac{dQ_s}{d\phi_s} = \frac{K_s\epsilon_0}{x_d}.$$

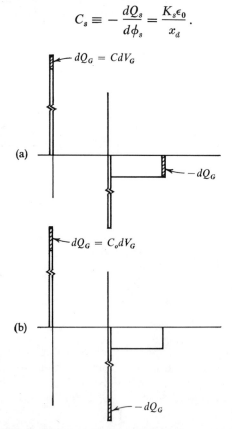

Fig. 9.9 Illustration of the variation of charge distribution in an MOS structure at (a) high frequencies; (b) low frequencies.

The Ideal MIS (or MOS) Structure

This is, in fact, what happens when the measurement frequency is high. If however the measurement frequency is low enough that recombination-generation rates can keep up with the small signal variation, then the recombination-generation mechanism will lead to charge exchange with the inversion layer in step with the measurement signal. In that case, the capacitance measured will approach that of the oxide layer alone.

To understand this better, let us consider what happens when a positive voltage applied to an MOS structure is increased by a small amount. Instantaneously, as the voltage is increased, more negative charge is induced in the silicon. At high frequencies, holes will be pulled out of the p-type semiconductor and the width of the depletion region will increase slightly, as shown in Figure 9.9a. If however electron-hole pairs can be generated fast enough, i.e., before the voltage is reduced again, the generated holes will replenish the holes pulled out from the edge of the depletion region and the extra electrons will appear in the inversion layer. Thus the incremental negative charge brought about by the incremental increase in gate voltage will appear at the oxide-silicon interface, as illustrated in Figure 9.9b. As a result, the capacitance measured will be that of the oxide layer alone, C_o.

As a consequence, the capacitance-voltage characteristics of metal-oxide-semiconductor structures are frequency dependent. Experimental measurements on an MOS structure corresponding to the computations given in Figure 9.8 are shown in Figure 9.10, for various measurement

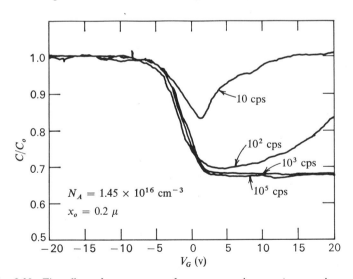

Fig. 9.10 The effect of measurement frequency on the capacitance-voltage characteristics of MOS structures.[5]

frequencies. In agreement with the above argument, it is evident that at low frequencies the capacitance approaches the oxide capacitance at the onset of strong inversion.

The transition from the "high-" to the "low-" frequency type capacitance-voltage characteristics can also be brought about at a given measurement frequency by an increase in the carrier recombination-generation rate, for instance by an increase in temperature or by illumination.[3]

Low-frequency type characteristics can also be observed at relatively high measurement frequencies due to two-dimensional effects, such as the existence of an inversion layer beyond the metal field plate.[6] This phenomenon will be discussed in Chapter 12. Also, in gate-controlled p-n junctions and in MOS surface field-effect transistors, where contact is made to the inversion layer through metallurgical junctions (see Figure 9.2), minority carriers can be supplied to the inversion layer through this external contact. Thus low-frequency type characteristics will prevail up to much higher frequencies because the supply of minority carriers is no longer dependent on a generation-recombination process.

c. Channel Conductance

Using the MOS surface field-effect transistor, structure (c) of Figure 9.2, an important complementary measurement can be performed. This measurement pertains to the conductance of an inversion layer or *channel* induced by a positive gate voltage (in the case of a p-type substrate). This conductance is given by

$$g = \frac{Z}{L} \int_0^{x_i} \sigma(x)\, dx \qquad (9.22)$$

where Z is the width of the conducting channel (in a direction normal to Figure 9.2), L is the distance of separation between the two n^+ regions, and $\sigma(x)$ is the conductivity of the inversion layer, given by

$$\sigma(x) = q\mu_n n(x). \qquad (9.23)$$

The point $x = x_i$ denotes the distance below the surface where the semiconductor is just intrinsic. Thus the channel conductance becomes

$$g = -\frac{Z}{L}\mu_n Q_n \qquad (9.24)$$

where Q_n is the charge density per unit area due to mobile carriers in the inversion layer. Combining Equation 9.24 with 9.10 yields

$$g = -\frac{Z}{L}\mu_n [Q_s - Q_B]. \qquad (9.25)$$

The Ideal MIS (or MOS) Structure

Combining Equation 9.25 with 9.16 and 9.21 yields

$$g = \frac{Z}{L}\mu_n C_o(V_G - V_T). \quad (9.26)$$

Thus, if the electron mobility μ_n is independent of the gate voltage, a straight line should result if conductance is plotted as a function of gate

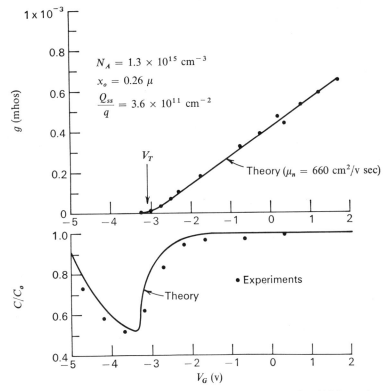

Fig. 9.11 The channel conductance and gate capacitance of an MOS transistor.[3]

voltage. Pertinent experimental results are shown in Figure 9.11 along with the corresponding capacitance-voltage measurements. Note that in accordance with the above argument, the conductance is indeed proportional to the quantity $(V_G - V_T)$. The voltage at which the conductance begins to increase is V_T, which explains why this voltage is called the turn-on voltage. Note that the onset of conductance closely corresponds to the rise in capacitance.

This simple derivation of the channel conductance applies only when the voltage drop V between the two n^+ regions is small in comparison to

($V_G - V_T$). When this condition is not met, as is often the case in practical operation of MOS transistors, the expression for the conductance along the surface takes on a more complicated form. This will be considered in detail in Chapter 11.

9.3 EFFECT OF WORK FUNCTION DIFFERENCE, CHARGES, AND STATES ON MOS CHARACTERISTICS

a. Work Function Difference

The electron energies at the Fermi level in the metal and in the semiconductor of an MOS structure will, in general, be different. Such an energy difference is usually expressed as a difference in _work functions_, which is the energy required to remove an electron from the Fermi level in a given material to vacuum. When the metal of an MOS structure is shorted to the semiconductor, electrons will flow from the metal to the semiconductor or vice versa until a potential will be built up between the two which will counterbalance the difference in work functions. When equilibrium is reached, the Fermi level in the metal is lined up with the Fermi level in the semiconductor. Therefore, there will be an electrostatic potential variation from one region to the other, as illustrated in Figure 9.12a for the case of an aluminum/silicon dioxide/p-type silicon sample.

In this figure we also show the conduction band of the oxide. The energy required to move an electron from the metal Fermi level into the conduction band of the oxide is called the _metal-oxide barrier energy_. The energy required to move an electron from the silicon valence band to the conduction band of the oxide is the _silicon-oxide barrier energy_. These barrier energies which are related to the respective work functions modified by the presence of the oxide can be independently measured by illuminating MOS structures with light of increasing frequency until the photon energy will be high enough to excite electrons into the conduction band of the oxide, thereby resulting in electronic conduction between metal and semiconductor. The barrier energies in the MOS system are discussed further in Chapter 12.

To derive the effect of a work function difference on MOS characteristics, it is easiest to consider the condition in which just enough gate voltage is applied to counterbalance the work function difference and a _flat-band condition_ is maintained in the semiconductor, as illustrated in Figure 9.12b. The gate voltage required to bring about the flat-band condition is called the _flat-band voltage_ V_{FB}.

Effects on MOS Characteristics

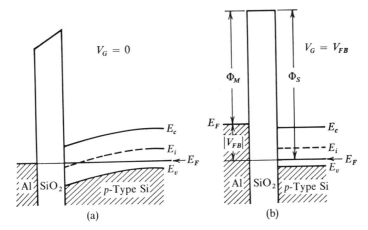

Fig. 9.12 The effect of metal-semiconductor work function difference on the potential distribution in an MOS structure.
(a) Conditions for $V_G = 0$.
(b) Flat-band condition.

The flat-band voltage is clearly the difference of the modified work functions indicated in Figure 9.12b,

$$V_{FB} = \Phi_M - \Phi_S \equiv \Phi_{MS}. \qquad (9.27)$$

We will discuss the role of the flat-band voltage on the MOS characteristics later in this section.

b. Charges in the Insulator

Consider a sheet charge per unit area Q within the insulator of a metal-insulator-semiconductor structure as shown in Figure 9.13. Under the conditions of zero gate voltage (Figure 9.13a), this sheet charge will induce an image charge partly in the metal and partly in the semiconductor. The resulting field distribution is indicated in the lower part of Figure 9.13a, neglecting work function differences and electrostatic potential variation in the semiconductor.

In order to bring about a flat-band condition (i.e., no charge induced in the semiconductor), we have to apply a negative voltage to the metal, as shown in Figure 9.13b. With increasing negative voltage, we are putting more negative charge on the metal and thereby shift the electric field distribution downwards until the electric field reaching the silicon surface becomes zero. Under this condition the area contained under the electric field distribution is the flat-band voltage V_{FB}. On the basis of this figure and Poisson's equation, V_{FB} is given by

$$V_{FB} = -\frac{xQ}{K_o\epsilon_0} = -\frac{x}{x_o}\frac{Q}{C_o}. \qquad (9.28)$$

Thus the flat-band voltage not only depends on the density of the sheet charge Q but also on its location within the insulator. When the sheet

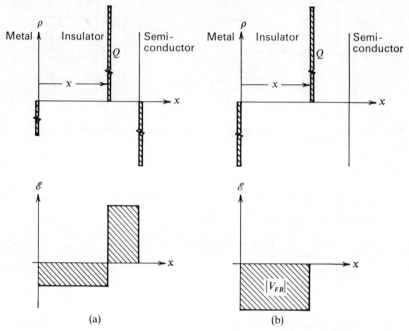

Fig. 9.13 The effect of a sheet charge within the insulator.
(a) Conditions for $V_G = 0$.
(b) Flat-band condition.

charge is next to the metal, it will induce no image charge in the silicon and, therefore, have no effect on the semiconductor surface. In the other extreme, when the sheet charge is located next to the semiconductor, it will exert its maximum influence, and lead to a flat-band voltage of

$$V_{FB} = -\frac{x_o Q}{K_o \epsilon_0} = -\frac{Q}{C_o}. \qquad (9.29)$$

We will see in Chapter 12 that there is indeed such a sheet charge associated with the silicon dioxide-silicon interface. Its density is denoted by Q_{ss}.

The more general case of an arbitrary space-charge distribution within the insulator is illustrated in Figure 9.14 for the flat-band condition. It

Effects on MOS Characteristics

can be shown by superposition of individual elements of such a charge distribution that the corresponding flat-band voltage is given by

$$V_{FB} = -\frac{1}{C_o} \int_0^{x_o} \frac{x}{x_o} \rho(x)\, dx. \tag{9.30}$$

Such space charges in the insulator can be due to several causes such as ionic contamination or traps ionized as a result of exposure to ionizing radiation. We will discuss both of these in detail in Chapter 12.

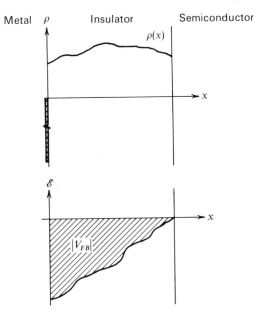

Fig. 9.14 The effect of an arbitrary space-charge distribution within the insulator. (Flat-band condition.)

Space charge or interface charge in the insulator, and a metal-semiconductor work function difference will both lead to a translation of the flat-band point from $V_G = 0$ along the voltage axis. Experimental observation of the flat-band voltage shift is illustrated in Figure 9.15. Note that the entire capacitance-voltage characteristic is translated in a parallel manner along the voltage axis by the same amount as the flat-band point. The *total* translation of the flat-band voltage or of any other well defined point on the capacitance-voltage (or, alternatively, the channel-conductance versus gate voltage) characteristic will then be given by

$$V_{FB} = \Phi_{MS} - \frac{Q_{ss}}{C_o} - \frac{1}{C_o}\int_0^{x_o} \frac{x}{x_o}\rho(x)\,dx. \tag{9.31}$$

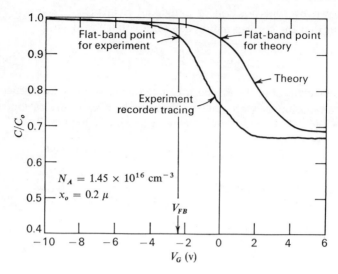

Fig. 9.15 The combined effects of metal-semiconductor work function difference and charges within the oxide on the capacitance-voltage characteristics of MOS structures.[3]

Thus direct comparison between theory and experiment is possible by plotting experiments as a function of $(V_G - V_{FB})$ rather than as a function of V_G alone. Experimental data are presented in this manner for thermally oxidized silicon in Figures 9.16 and 9.17. These data show the effect of substrate impurity concentration and of oxide thickness on the shape of the characteristics in comparison with the exact theoretical computations. It is evident that a uniform translation along the voltage axis brings about a very good agreement between theory and experiment in all cases.

c. Surface States

In Figure 9.3 we have shown how the application of a potential to the gate of an MOS structure will result in a movement of the energy bands relative to the Fermi level near the surface. If there are states within the forbidden gap concentrated at the surface, such as the surface recombination-generation centers discussed in Chapter 5, the probability of the occupation of these surface states will change as a result of the variation in band bending. This will happen because as the bands are moved up and down at the surface, the proximity of the energy level of the surface states to the Fermi level will change. This effect is illustrated in Figure 9.18 where we show the charge condition of surface states with energy level near the middle of the forbidden gap. As the surface is accumulated,

Effects on MOS Characteristics

these states will be lifted far above the Fermi level. Thus their probability of being occupied by electrons will be small and most of them will be unoccupied. As the surface is depleted and inverted, the states are pulled well below the Fermi level and their probability of occupation by electrons

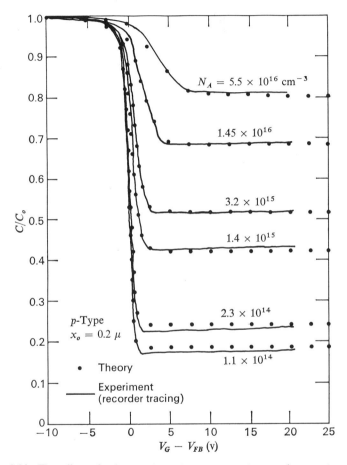

Fig. 9.16 The effect of substrate impurity concentration on the capacitance-voltage characteristics of MOS structures.[3]

will approach unity. Traditionally, such states whose charge can be readily exchanged with the semiconductor are called *fast surface states*.

Thus the charge in the fast surface states will vary with the band bending or surface potential ϕ_s. Owing to this change, the capacitance-voltage or conductance-voltage characteristics will be displaced from the theoretical characteristics *by an amount which itself varies with the surface potential.*

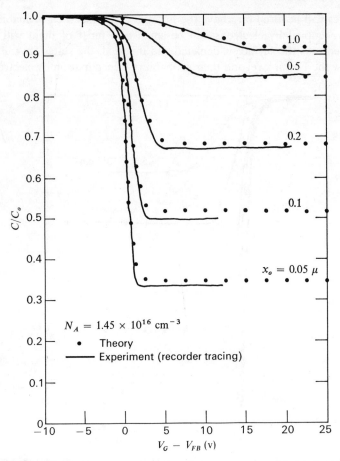

Fig. 9.17 The effect of oxide thickness on the capacitance-voltage characteristics of MOS structures.[3]

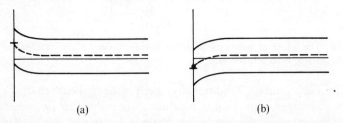

Fig. 9.18 The charge condition of a particular set of fast surface states with energy near the middle of the forbidden gap as a function of surface potential variation.
(a) Surface accumulated: states are unoccupied.
(b) Surface inverted: states occupied.

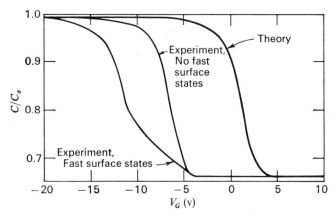

Fig. 9.19 The effect of fast surface states on the capacitance voltage characteristics of MOS structures.[7]

The result may appear either as steps or kinks in the characteristics or, for the case of a continuum of states in the forbidden gap, as a gradual distortion of the characteristics. An experimental illustration of the effect of fast surface states on MOS characteristics is shown in Figure 9.19.

READING REFERENCES

A more comprehensive treatment of the theory of semiconductor surfaces along with a bibliography of the earlier work on the subject can be found in A. Many, Y. Goldstein, and N. B. Grover, *Semiconductor Surfaces*, Wiley, 1965.

REFERENCES CITED

1. The MOS structure was first proposed as a voltage variable capacitor by J. L. Moll, "Variable Capacitance with Large Capacity Change," Wescon Convention Record, Part 3, p. 32 (1959); and by W. G. Pfann and C. G. B. Garrett, "Semiconductor Varactors Using Surface Space-Charge Layers," *Proc. IRE* (Correspondence), **47**, 2011 (1959). Its characteristics were then analyzed in detail by D. R. Frankl, "Some Effects of Material Parameters on the Design of Surface Space-Charge Varactors," *Solid-State Electron.*, **2**, 71 (1961); and by R. Lindner, "Semiconductor Surface Varactor," *Bell System Tech. J.*, **41**, 803 (1962). It was first employed in the study of thermally oxidized silicon surfaces by L. M. Terman, "An Investigation of Surface States at a Silicon/Silicon Dioxide Interface Employing Metal-Oxide-Silicon Diodes," *Solid-State Electron.*, **5**, 285 (1962); and by K. Lehovec and A. Slobodskoy, "Field Effect-Capacitance Analysis of Surface States on Silicon," *Phys. Stat. Solidi*, **3**, 447 (1963).

2. The theory of surface space-charge regions was developed by W. L. Brown, "N-type Surface Conductivity on P-type Germanium," *Phys. Rev.*, **91**, 518 (1953); C. G. B. Garrett and W. H. Brattain, "Physical Theory of Semiconductor Surfaces," *Phys.*

Rev., **99**, 376 (1955); R. H. Kingston and S. F. Neustadter, "Calculation of the Space Charge, Electric Field, and Free Carrier Concentration at the Surface of a Semiconductor," *J. Appl. Phys.*, **26**, 718 (1955).

3. A. S. Grove, B. E. Deal, E. H. Snow, and C. T. Sah, "Investigation of Thermally Oxidized Silicon Surfaces Using Metal-Oxide-Semiconductor Structures," *Solid-State Electron.*, **8**, 145 (1965).

4. C. E Young, "Extended Curves of the Space Charge, Electric Field, and Free Carrier Concentration at the Surface of a Semiconductor, and Curves of the Electrostatic Potential Inside a Semiconductor," *J. Appl. Phys.*, **32**, 329 (1961).

5. A. S. Grove, E. H. Snow, B. E. Deal, and C. T. Sah, "Simple Physical Model for the Space-Charge Capacitance of Metal-Oxide-Semiconductor Structures," *J. Appl. Phys.* **35**, 2458 (1964).

6. E. H. Nicollian and A. Goetzberger, "Lateral AC Current Flow Model for Metal-Insulator-Semiconductor Capacitors," *IEEE Trans. Electron Devices*, **ED-12**, 108 (1965); S. R. Hofstein and G. Warfield, "Physical Limitations on the Frequency Response of a Semiconductor Surface Inversion Layer," *Solid-State Electron.*, **8**, 321 (1965).

7. B. E. Deal, M. Sklar, A. S. Grove, and E. H. Snow, "Characteristics of the Surface-State Charge (Q_{ss}) of Thermally Oxidized Silicon," *J. Electrochem. Society*, **114**, (March 1967).

PROBLEMS

9.1 Verify the expression given for the capacitance of the surface space-charge region, Equation 9.19.

9.2 Derive an expression for the surface field required:
   (a) To make the surface exactly intrinsic.
   (b) To bring about strong inversion.

9.3 Calculate and plot:
   (a) the charge in the semiconductor,
   (b) the surface field, and
   (c) the surface potential
   at the onset of strong inversion, as a function of the impurity concentration in the semiconductor.

9.4 Derive an expression giving the change of the turn-on voltage of an MOS structure as a function of temperature.

9.5 (a) Derive a formula giving the ratio of the minimum high frequency capacitance of an MOS structure $C_{\min}$ to the oxide capacitance C_o as a function of oxide thickness and substrate impurity concentration.
   (b) The impurity concentration in the silicon in a region near the oxide-silicon interface may be different from the concentration deep in the substrate; for instance, due to redistribution during oxidation. In such a case, what impurity concentration would be inferred from the $C_{\min}/C_o$ ratio?

9.6 Exact calculations of the capacitance C of an MOS structure at flat band show it to be less than the capacitance of the oxide layer C_o (see, for example, Figure 9.8). Give a physical explanation for this fact.

Problems

9.7 Derive the transition frequency between "low" and "high" frequency type capacitance-voltage characteristics by equating the generation rate within the surface depletion region (see Chapter 5) to the charging current.

9.8 We have seen that the flat-band voltage V_{FB} depends on the metal-semiconductor work function difference as well as on the surface-state charge density. Devise a way by which these two factors can be determined from measurements of the flat-band voltage of MOS capacitors having various oxide thicknesses.

9.9 Calculate the change in flat-band voltage corresponding to:
   (a) A uniform positive charge distribution in the oxide.
   (b) A triangular distribution, which is high near the metal and zero near the silicon.
   (c) A triangular distribution which is high near the silicon and zero near the metal.
   Let the total density of ions be 10^{12} cm^{-2} in all three cases and consider a 0.2 μ thick oxide film.

9.10 Derive the charge per unit area Q_{st} in fast surface states as a function of surface potential for:
   (a) Single-level centers located at $E_t = E_i$, with density N_{st} (cm^{-2}).
   (b) Centers uniformly distributed in energy throughout the forbidden gap, with density D_{st} (cm^{-2} ev^{-1}).
   Assume that the surface states are acceptor type, i.e., negative when occupied by an electron, neutral otherwise.

9.11 Derive the capacitance-voltage characteristics of an MOS structure in the presence of single-level and uniformly distributed fast surface states as described in Problem 9.10.

TABLE 9.1 IMPORTANT FORMULAS FOR SURFACE SPACE-CHARGE REGIONS

		p-Type semiconductor	n-Type semiconductor
Charge per unit area in semiconductor	before strong inversion	$Q_s = -qN_A x_d$	$Q_s = qN_D x_d$
	after strong inversion	$Q_s = Q_n + Q_B$	$Q_s = Q_p + Q_B$
Surface potential at onset of strong inversion, *in thermal equilibrium*†		$\phi_s(\text{inv}) = 2\phi_{Fp} > 0$	$\phi_s(\text{inv}) = 2\phi_{Fn} < 0$
Maximum width of surface depletion region		$x_{d\max} = \sqrt{\dfrac{2K_s \epsilon_0 \phi_s(\text{inv})}{qN_A}}$	$x_{d\max} = \sqrt{\dfrac{2K_s \epsilon_0 |\phi_s(\text{inv})|}{qN_D}}$
Charge per unit area in depletion region, after strong inversion		$Q_B = -qN_A x_{d\max} = -\sqrt{2K_s \epsilon_0 qN_A \phi_s(\text{inv})}$	$Q_B = qN_D x_{d\max} = \sqrt{2K_s \epsilon_0 qN_D |\phi_s(\text{inv})|}$
Turn-on voltage of MIS structure		$V_T - V_{FB} = -\dfrac{Q_B}{C_o} + \phi_s(\text{inv}) > 0$	$V_T - V_{FB} = -\dfrac{Q_B}{C_o} + \phi_s(\text{inv}) < 0$
Conductance of inversion layer		$g = -\dfrac{Z}{L}\mu_n Q_n = \dfrac{Z}{L}\mu_n C_o (V_G - V_T)$ for $V_G > V_T$	$g = \dfrac{Z}{L}\mu_p Q_p = -\dfrac{Z}{L}\mu_p C_o (V_G - V_T)$ for $|V_G| > |V_T|$

† Under non-equilibrium conditions, $\phi_s(\text{inv}) = 2\phi_{Fp} + V_J$, and $\phi_s(\text{inv}) = 2\phi_{Fn} + V_J$ for p- and n-type semiconductors, respectively (see Chapter 10).

- **CHARACTERISTICS OF SURFACE SPACE-CHARGE REGIONS—NON-EQUILIBRIUM CASE**
- **THE GATE-CONTROLLED DIODE STRUCTURE**
- **RECOMBINATION-GENERATION IN THE SURFACE SPACE-CHARGE REGION**
- **FIELD-INDUCED JUNCTIONS AND CHANNEL CURRENTS**
- **SURFACE EFFECTS ON JUNCTION BREAKDOWN VOLTAGE**

10
Surface Effects on *p-n* Junctions

Perhaps the principal motivation for studying the properties of semiconductor surfaces is that surface effects can dominate the characteristics of *p-n* junctions and transistors. In Chapter 9 we saw how the metal-insulator-semiconductor capacitor structure can be employed to study the characteristics of surface space-charge regions. In this chapter we now show how surface space-charge regions can affect the characteristics of *p-n* junctions.

We begin with an extension of the theory of surface space-charge regions to non-equilibrium conditions such as are encountered in the vicinity of a biased *p-n* junction. We then develop the characteristics of the gate-controlled diode structure which takes the place of the metal-insulator-semiconductor structure as the principal experimental tool. We next consider the recombination-generation processes taking place in the surface space-charge region, and their influence on the current gain of transistors. Following this, we study breakdown phenomena associated with field-induced junctions and show how they can lead to catastrophic changes in both reverse currents of diodes and in current gain of transistors. Finally, we consider how surface fields can influence the breakdown voltage of planar junctions.

10.1 CHARACTERISTICS OF SURFACE SPACE-CHARGE REGIONS—NON-EQUILIBRIUM CASE[1]

Although the gate-controlled diode structure shown in Figure 10.1a refers to a diode having a *p*-type substrate, the following discussion is equally valid for a diode with an *n*-type substrate if appropriate changes in

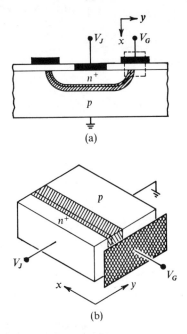

Fig. 10.1 (a) The gate-controlled diode structure.
(b) Idealized representation of the portion of the surface enclosed by the dashed frame in (a).[1]

the signs of the applied voltages are made. All voltages in this discussion are applied with the substrate held at ground potential. For simplicity, the *n*-type region is assumed to be much more heavily doped than the substrate. The condition of the surface of the substrate within the dashed frame in Figure 10.1a is first examined at different values of gate voltage V_G and reverse voltage V_R, in the absence of surface states or work function difference. This region is shown in a more idealized form in Figure 10.1b.

The idealized junction and the energy bands pertaining to thermal

Characteristics of Surface Space-Charge Regions

equilibrium conditions (i.e., when the applied junction voltage $V_J = 0$) are illustrated in Figure 10.2. In this energy band representation electron energy, as represented by the conduction and valence bands, is shown as a function of the two directions x and y corresponding to the axes of Figure 10.1. In the absence of surface fields the energy bands do not vary in the x direction. The only variation is in the y direction and is due to the built-in voltage ϕ_B between the n and p regions. This case is shown in Figure 10.2a. When a sufficiently large positive voltage $V_G > V_T$ is applied to the gate, the conduction band at the surface is brought close to the Fermi level and the surface of the p region becomes inverted. (V_T, the turn-on voltage, denotes the gate voltage necessary to cause inversion of the p surface.) The energy bands corresponding to this case are illustrated in Figure 10.2b. A *field-induced junction* now exists between the n-type inversion layer and the underlying p-type silicon. Since bias is not applied across this junction, it is in equilibrium and is characterized by the same Fermi level as the metallurgical junction. The surface space-charge region under these conditions is described by the equilibrium theory discussed in the previous chapter.

Two aspects of the equilibrium theory of surface space-charge regions should be recalled at this point. First, the total bending of the bands, as designated by the surface potential ϕ_s, is limited to some value less than the band gap of the semiconductor, i.e., about 1 volt for Si. More specifically, ϕ_s corresponding to strong inversion is given to a good approximation by $\phi_s(\text{inv}) = 2\phi_{Fp}$ where ϕ_{Fp} is the Fermi potential of the substrate. This value is the built-in voltage of the field-induced junction. Second, the width of the surface depletion region x_d first increases with increasing V_G and then reaches a maximum value $x_{d\max}$ when the surface becomes inverted. This width is the zero-bias depletion region width of the field-induced junction.

Figure 10.3 illustrates the non-equilibrium condition which exists when a reverse bias V_R is applied to the junction. Figure 10.3a shows the case in the absence of surface fields. In Figure 10.3b, a positive gate voltage is applied, but it is not large enough to invert the surface of the p-region. This is denoted by the condition $V_G < V_T(V_R)$, where $V_T(V_R)$ is the gate voltage necessary to cause inversion of the p region in the presence of a reverse bias V_R. *This voltage is larger than in the zero junction bias case.* This is because application of the reverse bias lowers the quasi-Fermi level for electrons so that even if the bands at the surface are bent as deeply as in the equilibrium case shown in Figure 10.2b, the band bending is still insufficient to bring the conduction band near enough to the quasi-Fermi level for electrons to cause inversion. As a result, the surface is only depleted.

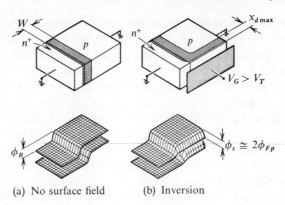

Fig. 10.2 p-n Junction under the influence of surface fields—equilibrium case ($V_J = 0$).[1]

In the case illustrated in Figure 10.3c, the voltage applied to the gate is large enough to overcome the influence of the reverse bias and an inversion layer is formed at the surface of the p region. In terms of the band diagram, the bands are now bent deeply enough to bring the conduction band near the quasi-Fermi level for electrons. Once an inversion layer is formed, it constitutes a region of high conductivity which is at essentially the same potential as the n region. The surface potential ϕ_s at the onset of strong inversion is now given, to a good approximation, by

$$\phi_s(\text{inv}) = V_R + 2\phi_{Fp}. \tag{10.1}$$

As in the equilibrium case, the surface depletion region reaches a maximum width $x_{d\max}$ at inversion. This width, however, is now a function of the reverse bias V_R, and is in fact the reverse-bias depletion region width of the field-induced junction formed between the n-type inversion layer and the underlying p region.

The most important features of the above discussion are summarized in Figure 10.4, where the charge distribution and energy bands of an inverted p-type substrate are shown as a function of the distance x from the surface, both for the equilibrium case (corresponding to Figure 10.2b), and for the reverse biased case (corresponding to Figure 10.3c).

The characteristics of the surface space-charge region will be given by the depletion approximation in a manner analogous to the equilibrium case considered in the previous chapter. The only difference is that the condition of the onset of strong inversion corresponding to the criterion $n_s = N_A$ will now be $\phi_s(\text{inv}) = V_J + 2\phi_{Fp}$ because of the splitting of the quasi-Fermi levels brought about by the applied bias. Thus the surface potential ϕ_s at the onset of strong inversion will be larger in the presence

Characteristics of Surface Space-Charge Regions

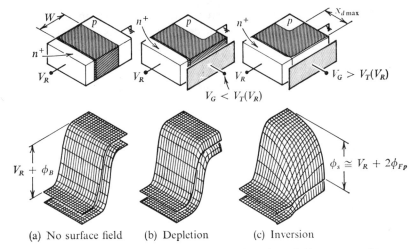

(a) No surface field (b) Depletion (c) Inversion

Fig. 10.3 p-n Junction under the influence of surface fields—reverse bias case $(V_J = V_R > 0)$.[1]

of a reverse bias V_R and smaller in the presence of a forward bias V_F applied to the p-n junction.

Otherwise, the description of the surface space-charge region will be unaltered. The electrostatic potential variation in the depletion region will be given by

$$\phi = \phi_s\left(1 - \frac{x}{x_d}\right)^2 \tag{10.2}$$

where the surface potential ϕ_s, which designates the total bending of the energy bands, is given by

$$\phi_s = \frac{qN_A x_d^2}{2K_s\epsilon_0}. \tag{10.3}$$

The maximum width of the depletion region will now be given by

$$x_{d\max} = \sqrt{\frac{2K_s\epsilon_0[V_J + 2\phi_{Fp}]}{qN_A}} \tag{10.4}$$

where we have used Equation 10.1 for $\phi_s(\text{inv})$.

The charge per unit area induced in the silicon prior to the onset of strong inversion is given by

$$Q_s = -qN_A x_d, \tag{10.5}$$

while after strong inversion it is given by

$$Q_s = Q_n + Q_B \tag{10.6}$$

where

$$Q_B \equiv -qN_A x_{d\max}. \tag{10.7}$$

Calculations based on these relationships are shown in Figure 10.5 where we show the surface depletion region width x_d as a function of the charge induced in the silicon Q_s for various values of applied junction bias. (All calculations in this chapter are for silicon at 300°K unless

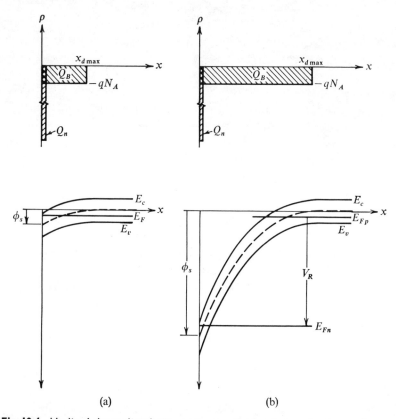

Fig. 10.4 Idealized charge distribution and energy band variation of an inverted p-region in a direction normal to the surface, some distance away from the junction.[1]
(a) Equilibrium case ($V_J = 0$), also shown in Fig. 10.2b.
(b) Reverse bias case ($V_R \simeq 5$ v), also shown in Fig. 10.3c.

otherwise specified.) The lines are based on the two limits of the depletion approximation; the points are based on more exact computer calculations. In Figure 10.6 we show the maximum width of the surface depletion region $x_{d\max}$ as a function of the applied reverse bias, for various substrate impurity concentrations. Note the similarity between this figure and Figure 6.9 which shows depletion region width versus applied reverse bias for a one-sided step junction.

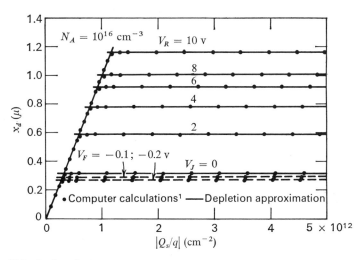

Fig. 10.5 Surface depletion region width versus charge induced in the semiconductor under various junction bias conditions.

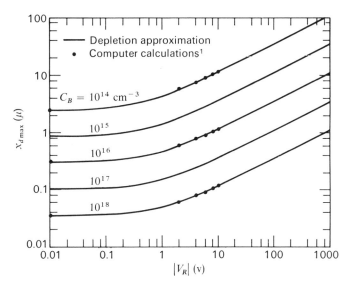

Fig. 10.6 The maximum width of the surface depletion region as a function of the magnitude of the junction reverse voltage. The parameter is substrate impurity concentration.

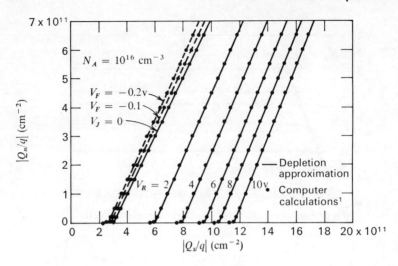

Fig. 10.7 The magnitude of the charge of minority carriers within the inversion layer as a function of the magnitude of the total charge induced within the semiconductor, under various junction bias conditions.

The charge due to electrons in the inversion layer is shown as a function of the total charge induced in the silicon for various values of forward and reverse bias in Figure 10.7.

10.2 GATE-CONTROLLED DIODE STRUCTURE[2]

The relationship between the potential applied to the gate and the characteristics of the surface space charge can be calculated exactly as in the case of the metal-insulator-semiconductor structure. Thus the gate voltage is given by

$$V_G = V_o + \phi_s \tag{10.8}$$

or

$$V_G = -\frac{Q_s}{C_o} + \phi_s, \tag{10.9}$$

and the gate-to-substrate capacitance per unit area, as before, will be given by the series combination of the oxide and semiconductor capacitances,

$$C = \frac{1}{\dfrac{1}{C_o} + \dfrac{1}{C_s}}. \tag{10.10}$$

Gate-Controlled Diode Structure

Calculations of the gate-to-substrate capacitance-voltage characteristics of a gate-controlled diode are shown in Figure 10.8. Note that prior to the onset of strong inversion the characteristics follow the depletion approximation; then, at the onset of strong inversion, in the low frequency case shown here, the capacitance rapidly increases to the oxide capacitance. (As discussed in Chapter 9, in the case of gate-controlled diodes, minority carriers can be supplied to the inversion layer through the external circuit.

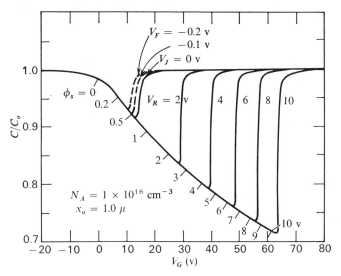

Fig. 10.8 Theoretical low frequency gate-to-substrate capacitance-voltage characteristics of a gate-controlled diode structure under various applied junction bias conditions. Values of surface potential ϕ_s are indicated.[1]

For this reason, the "low frequency" type characteristics will be observed to much higher frequencies than in the case of MOS capacitors.)

The point of onset of strong inversion is considerably displaced from the equilibrium case as the reverse bias is increased. Its location, the *turn-on voltage*, can be calculated from Equation 10.9 by substituting the conditions for strong inversion, $Q_s = Q_B$ and $\phi_s = 2\phi_{Fp} + V_J$. This yields

$$V_T = -\frac{Q_B}{C_o} + 2\phi_{Fp} + V_J. \tag{10.11}$$

Because Q_B itself is a function of V_J, the change in turn-on voltage with applied junction bias V_J can be considerably larger than the junction bias itself. This is borne out by the characteristics shown in Figure 10.8.

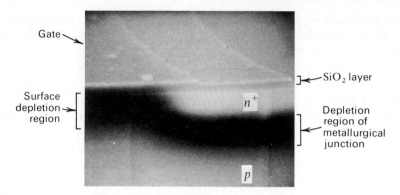

Fig. 10.9 Direct observation of the surface space-charge region of a reverse biased gate-controlled n^+p diode, cleaved through the junction.[3]

As in the case of the simple metal-insulator-semiconductor capacitor structure, the effect of a work function difference or of charges in the oxide will be a displacement of the characteristics along the voltage axis by the flat-band voltage.

The theory of surface space-charge regions under non-equilibrium conditions has been verified by extensive electrical measurements on gate-controlled diodes.[1] Direct observations of the shape of surface depletion regions have also tended to verify the qualitative features of this theory. An example of such an observation made by a scanning electron microprobe,[3] is shown in Figure 10.9.

10.3 RECOMBINATION-GENERATION PROCESSES IN THE SURFACE SPACE-CHARGE REGION

We have seen in Chapter 6 that the room temperature reverse current of silicon p-n junctions is due to electron-hole pairs generated through the action of recombination-generation centers within the depletion region. Thus we expect that the magnitude of the reverse current depends on the total number of such centers included within the junction depletion region. With this in mind, let us now consider the reverse current versus gate voltage characteristics of a gate-controlled n^+p junction depicted in Figure 10.10.

When the surface under the gate is *accumulated*, only those centers which are within the depletion region of the metallurgical p-n junction

Recombination-Generation Processes

contribute to the generation current (Figure 10.10a). When the surface under the gate is *inverted*, centers within the surface depletion region, i.e., the depletion region of the field-induced junction, also contribute to the generation current which is, therefore, larger than in the first case (Figure 10.10c). This contribution to the generation current is related to the width of the surface depletion region x_d. Thus, when the surface is depleted and x_d is increasing with increasing gate voltage, this current

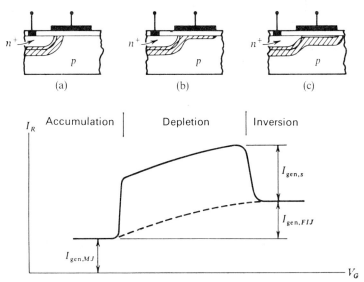

Fig. 10.10 Illustration of the effect of variation in the nature of the surface space-charge region on the reverse current of an n^+p diode at a fixed reverse voltage.[1]

component increases as indicated by the dashed line in Figure 10.10. Once the surface is inverted, x_d reaches its maximum value and hence there will be no further increase in this current component. However, while the surface is *depleted*, recombination-generation centers at the oxide-silicon interface provide yet another contribution to the total generation current. This contribution will result in a peak in the reverse current versus gate voltage characteristics (Figure 10.10b).

This picture is given verification by the measurements shown in Figure 10.11. Here we show both the reverse current and the gate-to-substrate capacitance as a function of gate voltage for various values of reverse voltage as a parameter. This figure shows that the reverse current increases at the same gate voltage regardless of the value of the reverse bias, corresponding to the depletion of the surface. The current decreases again, although not to its original value, when the surface becomes inverted as

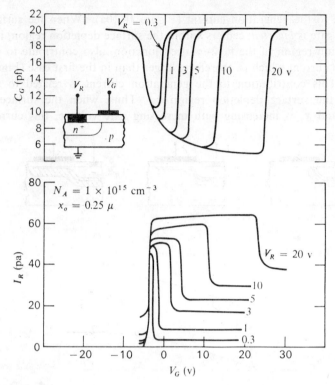

Fig. 10.11 Reverse current as a function of gate voltage, with junction reverse voltage V_R as parameter. Also shown are the corresponding gate-to-substrate capacitance measurements.

verified by the observation that the point of decrease in current coincides closely with the point of the capacitance rise.

Thus the generation current may consist of one or more of the following three components, depending on the nature of the surface space-charge region:

$$I_{\text{gen},MJ} = qU_{MJ}WA_{MJ} \qquad (10.12)$$

which corresponds to generation within the depletion region of the metallurgical junction;

$$I_{\text{gen},FIJ} = qU_{FIJ}x_{d\max}A_s \qquad (10.13)$$

which corresponds to generation within the depletion region of the field-induced junction. This equation applies only after the surface is inverted. When the surface is depleted, but not inverted, $x_{d\max}$ is replaced by x_d.

Recombination-Generation Processes

Finally,
$$I_{gen,s} = qU_s A_s \tag{10.14}$$

which corresponds to the surface generation component.

In the above equations, A_{MJ} denotes the area of the metallurgical junction and A_s the area of the substrate under the gate; U_{MJ} and U_{FIJ} are the carrier generation rates per unit volume in the reverse biased depletion regions of the metallurgical and field-induced junctions, respectively. U_s is the carrier generation rate per unit surface area at the oxide-silicon interface when this interface is completely depleted of both electrons and holes, i.e., when the electron and hole concentrations at the surface n_s and p_s are much smaller than the intrinsic carrier concentration n_i (a condition that is met when the junction is reverse biased).

We saw in Chapter 6 that the generation rate per unit volume in a reverse-biased depletion region is given by

$$U = \frac{1}{2}\frac{n_i}{\tau_o} \tag{10.15}$$

with $\tau_o = 1/\sigma v_{th} N_t$ for centers with energy level $E_t = E_i$. Here N_t is the concentration (per unit volume) of bulk recombination-generation centers, and σ is their capture cross section, assumed to be the same for electrons and holes for simplicity.

The carrier generation rate of a completely depleted surface per unit area is given by

$$U_s = \tfrac{1}{2} n_i s_o \tag{10.16}$$

with $s_o = \sigma v_{th} N_{st}$ on the basis of Equation 5.50, for centers with energy level $E_t = E_i$ where N_{st} is the density (per unit area) of surface recombination-generation centers and σ is their capture cross section, assumed to be the same for electrons and holes.

Thus the three current components become

$$I_{gen,MJ} = \tfrac{1}{2} q \frac{n_i}{\tau_{oMJ}} W A_{MJ}, \tag{10.17}$$

$$I_{gen,FIJ} = \tfrac{1}{2} q \frac{n_i}{\tau_{oFIJ}} x_{d\max} A_s, \tag{10.18}$$

and
$$I_{gen,s} = \tfrac{1}{2} q n_i s_o A_s. \tag{10.19}$$

The assumed condition of $E_t = E_i$ is in good agreement with experimental measurements of the temperature dependences of the three current components,[1] all of which have the same temperature dependence as n_i.

However, it is important to note that a *continuum* of states distributed in energy would lead to similar results since only those states that are within approximately one kT from the middle of the energy gap are efficient generation centers. For *uniform distributions* of states of density D_t (cm^{-3} ev^{-1}) and D_{st} (cm^{-2} ev^{-1}) in the bulk and at the surface, respectively, the effective lifetime τ_o is

$$\tau_o = \frac{1}{\sigma v_{th}(\pi k T D_t)}, \quad (10.20)$$

and the surface recombination velocity s_o is

$$s_o = \sigma v_{th}(\pi k T D_{st}). \quad (10.21)$$

Such states would also lead to current components which have the same temperature dependence as n_i. Thus single-level states located at energy E_i and a continuum of states will have the same effect.

It is evident from Equations 10.17, 10.18, and 10.19 that, whereas both bulk generation current components should depend on the magnitude of the reverse bias through W and $x_{d\max}$, the surface generation current should be independent of the reverse bias. The measurements shown in Figure 10.11 approximately bear out this conclusion.

Typical values of s_o on thermally oxidized silicon surfaces are of the order of 1 to 10 cm/sec. Typical values of τ_o are of the order of 1 to 10 μsec. For junction and surface areas of $\sim 10^{-3}$ cm^2, these magnitudes correspond to reverse current variations of the order of tens of picoamps as shown in Figure 10.11. Although these currents are small enough to be almost inconsequential from a practical standpoint, their significance becomes great under forward-bias conditions.

We have seen in Chapter 6 for the case of bulk current components that the recombination current in the forward-bias condition will be given approximately by

$$I_{\text{rec}} \cong I_{\text{gen}} e^{q|V_F|/2kT}. \quad (10.22)$$

A similar relationship also holds approximately for the field-induced junction and the surface components. Thus the maximum recombination current will be given approximately by

$$I_{\text{rec}} = \tfrac{1}{2} q n_i \left[\frac{W}{\tau_{oMJ}} + \left(\frac{x_{d\max}}{\tau_{oFIJ}} + s_o \right) \frac{A_s}{A_{MJ}} \right] e^{q|V_F|/2kT} A_{MJ}. \quad (10.23)$$

This expression shows that the recombination current under forward bias will be affected by the field-induced junction and the surface components *in the same percentage manner* as the generation current under reverse bias. If, for example, the reverse current varies by a factor of two,

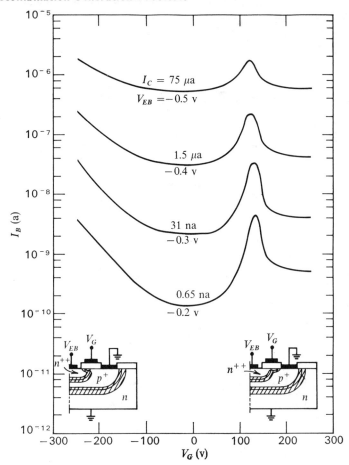

Fig. 10.12 Base current as a function of gate voltage, with emitter-base forward bias as parameter, for the transistor shown in the inset.[4] Surface concentration of base region is ~2.5 × 10^{18} cm^{-3}.

the forward recombination current will also vary by approximately a factor of two. Since, in the case of a transistor, all of the recombination current will appear as base current, this means that the common-emitter current gain h_{FE} of a transistor will vary in the same proportion. This effect is of extreme importance in controlling the current gain of transistors, especially at relatively low collector-current levels.†

† Recall that because recombination currents increase with the factor $\exp q|V_F|/2kT$, whereas the collector current increases with the factor $\exp q|V_F|/kT$, the recombination current components will become less and less important as the current level is increased. (See Chapter 7.)

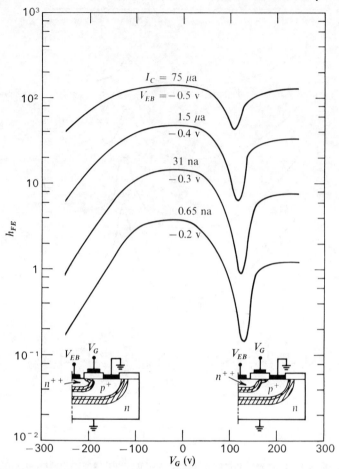

Fig. 10.13 Current gain $h_{FE} \equiv I_C/I_B$ as a function of gate voltage, corresponding to Fig. 10.12.

Experimental measurements[4] on a transistor whose emitter-base junction is controlled by a field plate are shown in Figure 10.12 where we show the base current as a function of gate voltage, for various values of emitter-base forward bias. The corresponding collector currents, which did not vary with gate voltage, are also indicated. As the gate voltage varies from negative to positive values, the base current goes through a peak and then it decreases again although not to its original value. The corresponding variation in the surface space-charge region associated with the emitter-base junction is indicated in the insets. The current gain $h_{FE} \equiv I_C/I_B$ is shown as a function of gate voltage in Figure 10.13. Where the base current had a peak, h_{FE} has a dip. After that it recovers partially, but not to its original value. These phenomena are in general agreement with the above discussion. (For the moment we will ignore the variation of base current for large negative voltages.)

10.4 FIELD-INDUCED JUNCTIONS AND CHANNEL CURRENTS

Let us now focus our attention in further detail on the case when the surface of the substrate is inverted so that a field-induced junction exists in parallel with the metallurgical junction, as illustrated in Figure 10.14.

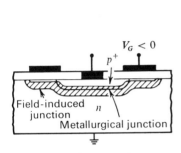

Fig. 10.14 Field-induced junction in parallel with a metallurgical junction.

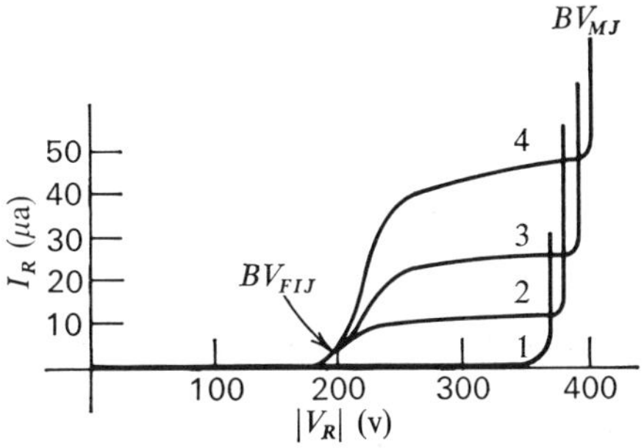

Fig. 10.15 Illustration of the consequence of the breakdown of the field-induced junction. Increasing numbers correspond to increasingly negative gate voltages.

This field-induced junction will have a breakdown voltage of its own. In particular, in many cases the field-induced junction will break down at some voltage which is lower than the breakdown voltage of the metallurgical junction. The reverse current-voltage characteristics in such a case are illustrated in Figure 10.15. When the reverse voltage exceeds the breakdown voltage of the field-induced junction BV_{FIJ}, current begins to flow.

This current flows along the inversion layer to the diffused region, and it will *saturate* upon further increase in reverse voltage.† This type of current-voltage characteristic is often referred to as a *channel characteristic*. The level at which the reverse current saturates depends on the conductance of the inversion layer. The larger this conductance, the higher the saturation current level, as shown in Figure 10.15.

† Such a phenomenon of current saturation is generally associated with current flow through conducting channels. It is discussed in Chapters 8 and 11.

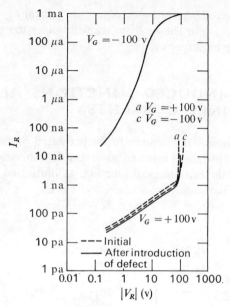

Fig. 10.16 The effect of the introduction of a low-breakdown producing defect on the reverse I-V characteristics of a gate-controlled p^+n diode, such as shown in Fig. 10.14.[5]

If there is a defect present within the field-induced junction shown in Figure 10.14 which lowers the breakdown voltage of the field-induced junction,† a large excess reverse current will be observed even at relatively low reverse biases. This effect is demonstrated in Figure 10.16 where we show the reverse current-voltage characteristics on log-log scales in order to illustrate the typical magnitudes involved.[5] The dashed lines show the

† Such a defect may be analogous to those which lead to the "soft" reverse current-voltage characteristics discussed in Chapter 6.

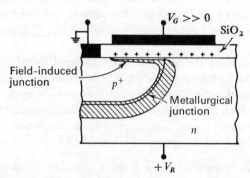

Fig. 10.17 Field-induced junction formed over the heavily doped region.

Field-Induced Junctions and Channel Currents

initial reverse current-voltage characteristics of a p^+n gate-controlled diode for $V_G = \pm 100$ volts. With the exception of a slight variation in the breakdown voltage, there is very little effect of gate voltage on the characteristics. A *defect* was then introduced under the gate by an electrostatic discharge across the gate. The resulting characteristics for $V_G = \pm 100$ volts are shown by the solid lines. When $V_G = +100$ volts, no field-induced junction exists and, therefore, the defect has no influence on the junction characteristic. When $V_G = -100$ volts, however, a field-induced junction does exist and a huge excess current flows due to the presence of the defect in the field-induced junction.

A conceptually very similar case involving a low value of the breakdown voltage of the field-induced junction is encountered if the field-induced junction is formed over the heavily doped region of the junction rather than over the substrate, as illustrated in Figure 10.17. Because the charges present in the silicon dioxide layer generally tend to be positive (see Chapter 12), this phenomenon is more likely to be encountered in practice over p^+ regions than over n^+ regions.

Because the field-induced junction is formed over material of high

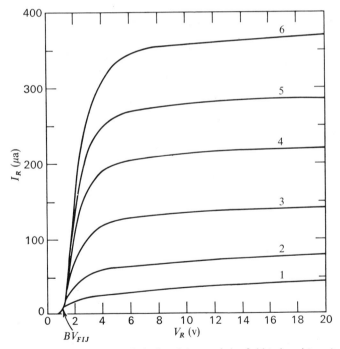

Fig. 10.18 The consequence of the breakdown of the field-induced junction shown in Fig. 10.17. Increasing numbers correspond to increasingly positive gate voltages.[6]

doping concentration, BV_{FIJ} will be low and the channel current will commence at a small value of the reverse voltage. An experimentally observed[6] set of such reverse current-voltage characteristics is shown in Figure 10.18 for increasingly positive gate voltages. Note that channel current begins to flow when the reverse voltage exceeds BV_{FIJ}, which in this case is approximately 1 volt.

The breakdown characteristics of field-induced junctions, i.e., the characteristics of channel currents well below saturation, are strongly affected by the surface concentration of the p^+ region. At low surface concentrations the breakdown mechanism is avalanche breakdown; at high surface concentration it is Zener or tunneling breakdown. Zener breakdown has the interesting property of being approximately symmetrical near the zero-bias point, i.e., it leads to a large excess current also flowing in the forward direction. The forward and reverse current-voltage characteristics of a

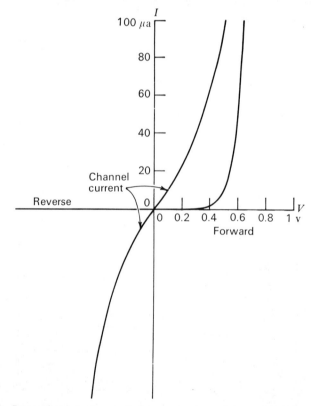

Fig. 10.19 Forward and reverse channel current-voltage characteristics of a p^+n junction with high boron surface concentration ($C_S = 2 \times 10^{19}$ cm^{-3}). Also shown are the original diode characteristics.[6]

Field-Induced Junctions and Channel Currents

field-induced junction formed over a p^+ region of high surface concentration are shown in Figure 10.19. These characteristics are evidently quite symmetrical.

Such large excess forward currents are exceedingly important in the case of the emitter-base junction of transistors. Because such excess currents do not contribute to transistor action, they will appear as base current.[4] This is illustrated in Figure 10.20 where we show the base current as a function of gate voltage of a transistor similar to that used in Fig. 10.12 with the exception that the surface concentration of the base region is higher than in the previous case. It is evident that in addition to the peak

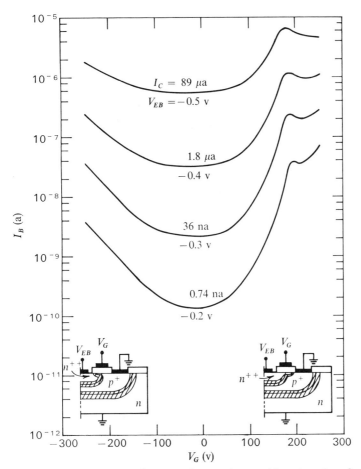

Fig. 10.20 Base current as a function of gate voltage, with emitter-base forward bias as parameter, for the transistor shown in the inset.[4] Surface concentration of the base region is $\sim 4 \times 10^{18}$ cm^{-3}.

due to surface recombination, the base current drastically increases for both sufficiently large negative and positive gate voltages. As indicated in the insets, these conditions correspond to the formation and Zener breakdown of field-induced junctions over the emitter and the base regions, respectively.

The drastic effect of such breakdown currents on the current gain of the transistor is shown in Figure 10.21. Note that this phenomenon can influence the current gain to much higher collector-current levels than surface recombination.

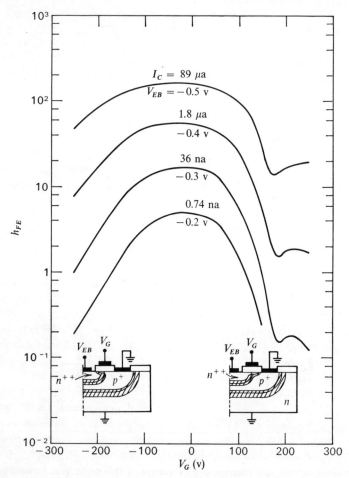

Fig. 10.21 Current gain $h_{FE} \equiv I_C/I_B$ as a function of gate voltage, corresponding to Fig. 10.20.

10.5 SURFACE EFFECTS ON JUNCTION BREAKDOWN VOLTAGE

In the preceding discussions we saw that the shape of the depletion region near the surface can be varied between the two extremes of forming a field-induced junction on the heavily doped side or on the substrate. Accordingly, the field distribution near the surface will also vary. Under many conditions the electric field near the surface will be higher than in the bulk and, therefore, breakdown will occur near the surface at a lower reverse bias than that corresponding to the breakdown voltage in the bulk.

The type of variation expected is illustrated in Figure 10.22 for an n^+p diode. For a very large negative gate voltage, a field-induced junction will be formed over the n^+ region (Figure 10.22a). Near the *"corner" region* the depletion region will be relatively narrow and, therefore, the electric field intensity there will be high. (In this figure the arrows designate only the *direction* of the electric field.) Thus junction breakdown will occur in this region at a relatively low value of the reverse bias. In the other extreme, shown in Figure 10.22c, the gate voltage is the same as the junction voltage. A surface depletion region of approximately the same width as the junction depletion region will be formed and, therefore, the electric field intensity near the corner will be reduced below even that obtained in an undisturbed planar *p-n* junction (i.e., with no surface fields). We would then expect the breakdown voltage to be high. In an intermediate case, for instance, when the gate is kept at the same potential as the substrate, the field near the corner region will be somewhat increased over the undisturbed case as shown in Figure 10.22b.

Experimental observations of the breakdown voltage of n^+p diodes as a function of the gate voltage are shown in Figure 10.23 for three different substrate impurity concentrations. In each case the breakdown voltage starts out from a relatively low value at negative gate voltages, increases as the gate voltage is made more positive and approaches the *plane* rather than *planar* value of the breakdown voltage in each case. The gate voltage that must be applied in order to approach the plane breakdown voltage values is approximately that required to bring about the formation of a field-induced junction over the substrate, i.e., the turn-on voltage. Calculated turn-on voltage values are designated by the cross-hatched line segments intersecting the data points. (It should be kept in mind that the turn-on voltage is a function of the reverse voltage. This was taken into account in the calculation of these line segments.)

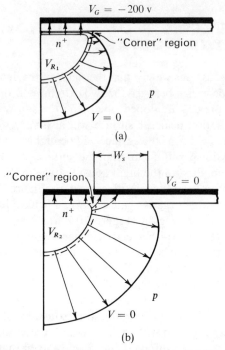

Fig. 10.22 Illustration of the effect of gate voltage on the shape of the depletion region and the breakdown voltage. V_{R_1}, V_{R_2} and V_{R_3} correspond to the breakdown voltages under the indicated gate-voltages.[7]

In the intermediate range, the breakdown voltage variation follows

$$BV = mV_G + \text{constant} \tag{10.24}$$

where $m \simeq 1$.

To explain this type of variation, we consider the potential distribution within the depletion region of an idealized gate-controlled diode calculated from numerical solution of Poisson's and Laplace's equations in the depletion region and in the oxide, respectively, shown in Figure 10.24. In all four cases the electric field intensity in the corner is evidently larger than elsewhere in the depletion region. It might be expected that, if $x_o \ll W$, this electric field will be more closely related to the field across the oxide $(V_R - V_G)/x_o$ than to the field across the depletion region. Thus the electric field at the corner is given by

$$\mathscr{E}_{\text{corner}} = \alpha \frac{V_R - V_G}{x_o} \tag{10.25}$$

where α is some geometric correction factor. It then follows that the condition of breakdown $\mathscr{E}_{\text{corner}} = \mathscr{E}_{\text{crit}}$ will lead to a relationship of the

Junction Breakdown Voltage

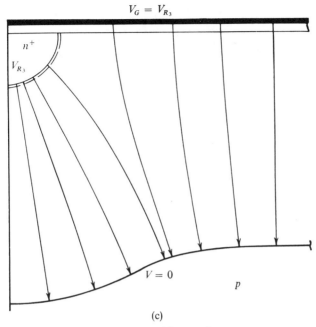

(c)

Fig. 10.22 (Continued)

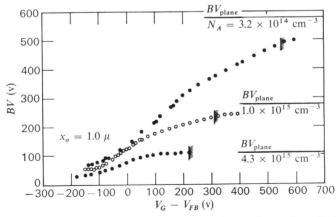

Fig. 10.23 The effect of gate voltage on the breakdown voltage of n^+p diodes. Cross-hatched line segments denote onset of inversion of substrate.[7]

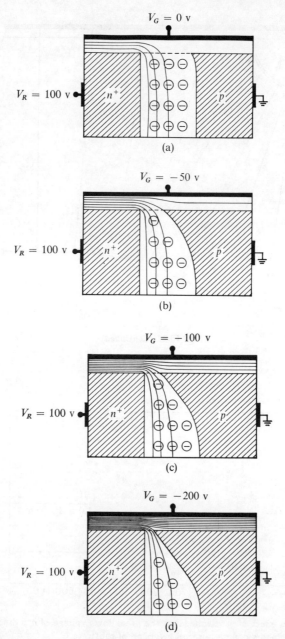

Fig. 10.24 The potential distribution in the depletion region of an idealized gate-controlled diode. Lines designate equipotentials 25 v apart. Results of numerical integration for $N_A = 2.5 \times 10^{15}$ cm^{-3}, $x_o = 1$ μ.[7]

same form as Equation 10.24, i.e.,

$$BV = V_G + \frac{\mathscr{E}_{\text{crit}} x_o}{\alpha}. \tag{10.26}$$

This relationship is in reasonable agreement with the experimental observations as well as with the corresponding estimates based on the numerical calculations.

The electric field near the corner is related to the electric field across the oxide only so long as the oxide thickness is much smaller than the depletion region width. As the substrate impurity concentration or the oxide thickness is increased, there will be an increasing deviation from this condition. This is borne out by the experimental data shown in Figure 10.23: as the impurity concentration is increased, the effect of gate voltage on the breakdown voltage becomes increasingly smaller.

REFERENCES CITED

1. A. S. Grove and D. J. Fitzgerald, "Surface Effects on *P-N* Junctions—Characteristics of Surface Space-Charge Regions under Non-Equilibrium Conditions," *Solid-State Electron.*, **9,** 783 (1966).
2. The field plate or gate-controlled *p-n* junction was employed in the study of germanium surface properties by J. H. Forster and H. S. Veloric, "Effect of Variations in Surface Potential on Junction Characteristics," *J. Appl. Phys.*, **30,** 906 (1959). This structure was used to study surface effects on silicon *p-n* junctions by C. T. Sah, "Effect of Surface Recombination and Channel on *P-N* Junction and Transistor Characteristics," *IRE Trans. Electron Devices*, **ED-9,** 94 (1962).
3. N. McDonald and T. Everhart, unpublished.
4. V. G. K. Reddi and C. A. Bittmann, "Second Quarterly Report on Micropower Functional Electronic Blocks," Contract AF 33 (615)-3010 (1966).
5. D. J. Fitzgerald and A. S. Grove, "Mechanisms of Channel Current Formation in Silicon *P-N* Junctions," *Physics of Failure in Electronics*, Volume 4, p. 315, Rome Air Development Center (1966).
6. A. S. Grove and D. J. Fitzgerald, "The Origin of Channel Currents Associated with *P*+ Regions in Silicon," *IEEE Trans. Electron Devices*, **ED-12,** 619 (1965).
7. A. S. Grove, O. Leistiko, and W. W. Hooper, "Effect of Surface Fields on the Breakdown Voltage of Planar Silicon *P-N* Junctions," *IEEE Trans. Electron Devices* **ED-14** , 157 (March, 1967).

PROBLEMS

10.1 Calculate and plot the turn-on voltage V_T as a function of reverse bias V_R for two gate-controlled diodes having substrate impurity concentrations of 10^{14} and 10^{16} cm^{-3}, respectively, and an oxide thickness of 1 μ. Take $V_{FB} = 0$.

10.2 A gate-controlled p^+n diode has a metallurgical junction area of 10^{-3} cm². The gate overlaps the n-region over an area of 10^{-3} cm². The substrate impurity concentration is 10^{16} cm⁻³, the junction depth is 5 μ, the oxide thickness is 0.2 μ, the lifetime $\tau = 1$ μsec, the surface recombination velocity $s_o = 5$ cm/sec. The flat-band voltage is $V_{FB} = -2$ v. Calculate:
   (a) The gate voltages at which the surface of the substrate is intrinsic, and is strongly inverted, respectively, for $V_J = 0$.
   (b) The room-temperature reverse current at $V_R = -1$ v, for $V_G = -20, 0,$ and $+20$ v.
   (c) The forward current at $V_F = 0.4$ v, at the same values of the gate voltage. Sketch the reverse and forward currents as a function of gate voltage.
   (d) The breakdown voltages at the above values of the gate voltage, and also in the absence of a gate.

10.3 For the diode of the previous problem, calculate the junction capacitance C_J at $V_G = 0$, and at $V_R = -2$ v for $V_G = -20, 0,$ and $+20$ v. Sketch C_J versus gate voltage. Also calculate the capacitance between the gate and the p^+ region, for $V_J = 0$, for the above values of the gate voltage.

10.4 Derive an expression giving the maximum variation of the current gain h_{FE} with gate voltage, for a transistor with a gate over the emitter-base junction, as a function of collector current. (Neglect excess currents due to the Zener mechanism.)

10.5 A defect under the gate of a gate-controlled diode lowers the breakdown voltage of the field-induced junction to 2 v. What gate voltage must be applied in order to bring about an excess current due to this defect? The substrate impurity concentration is 10^{16} cm⁻³, the oxide thickness is 1 μ. Take $V_{FB} = 0$.

10.6 Calculate and plot the surface field at the onset of strong inversion as a function of reverse bias for a gate-controlled diode which has a substrate impurity concentration of 10^{16} cm⁻³. Assuming the field-induced junction formed upon inversion is *plane*, estimate its breakdown voltage. Compare this breakdown voltage with that of a one-sided step junction formed within the same substrate.

10.7 A planar p^+n diode (no gate) contains 2×10^{12} positive charges/cm² within its oxide. The substrate impurity concentration is 10^{14} cm⁻³, the oxide thickness is 1 μ. Calculate its approximate breakdown voltage if the junction depth is (a) 5 μ and (b) 50 μ.

- PRINCIPLES OF OPERATION
- CHARACTERISTICS
- MODIFICATIONS OF THE SIMPLE THEORY
- OTHER TYPES OF SURFACE FIELD-EFFECT TRANSISTORS

11

Surface Field-Effect Transistors

The last semiconductor device that we consider here was perhaps the first one to be conceived. The principle of the surface field-effect transistor dates back to the early 1930's when Lillienfeld[1] in the United States and Heil[2] in England proposed to use the surface field-effect to achieve a solid-state amplifier. It was subsequently actively investigated by the Bell Laboratories group in the late 1940's.[3] The more or less accidental discovery of the bipolar transistor then gave new direction to solid-state device research and development for more than a decade.

The advent of thermally oxidized silicon brought with it an increase in the feasibility of fabricating the surface field-effect transistor. In 1960, Kahng and Atalla[4] used a thermally oxidized silicon structure in a surface field-effect transistor. The ensuing years brought about an exceedingly intense activity in this field. This activity, on the one hand, led to a high level of knowledge and understanding of the thermally oxidized silicon surface (see Chapter 12). On the other hand, this activity is responsible for the fact that the *MOS transistor*, i.e., the surface field-effect transistor using a thermally grown silicon dioxide layer, has become potentially the second most important device next to the bipolar transistor. In fact, in many integrated circuit applications, the MOS transistor may eventually become the more important one.

Although various kinds of surface field-effect transistors can be and have been fabricated, the discussion in this chapter deals with the

n-channel device which we have already considered in Chapter 9 in connection with studies of semiconductor surfaces. This device is illustrated in Figure 11.1. It consists of a p-type silicon substrate into which two n^+ regions, the *source* and the *drain*, are diffused. The region between the source and the drain is under the influence of a metal field plate or *gate*. If a large positive voltage is applied to the gate, the surface of the underlying p-type silicon can be inverted and a conductive n-type *channel* can be induced connecting source and drain. The conductivity of this channel can then be modulated by varying the gate voltage.

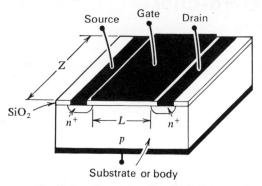

Fig. 11.1 n-Channel surface field-effect transistor.

We first consider the principles of the operation of such a surface field-effect transistor. Then we discuss some important characteristics of surface field-effect transistors: the current-voltage characteristics, the transconductance, and the gate leakage current. Finally, we consider the modifications of the simple theory. Insofar as possible, we attempt to maintain a parallelism between the treatment of the surface field-effect transistor and its sister device, the junction field-effect transistor discussed in Chapter 8.

11.1 PRINCIPLES OF OPERATION

Let us consider the situation when a large enough gate voltage is applied to induce an n-type inversion layer between the source and the drain regions, as shown in Figure 11.2. The cases of small and large drain voltages are considered separately. For small drain voltages, the channel induced between source and drain essentially behaves like a resistor. Its resistance, as shown in Chapter 9, is

$$R = -\frac{L}{Z\mu_n Q_n} \tag{11.1}$$

Principles of Operation

where Q_n is the charge density per unit surface area of electrons in the inversion layer. The magnitude of Q_n will depend on the silicon surface field; hence, it will depend on the potential difference between the gate and the inversion layer.

As the drain voltage is increased, the average potential difference from gate to the n-type inversion layer will decrease. As a result, Q_n will also decrease and the resistance of the channel increase. Thus the drain current versus drain voltage characteristic will begin to bend downward from the initial resistor line. This is evident in the experimental current-voltage

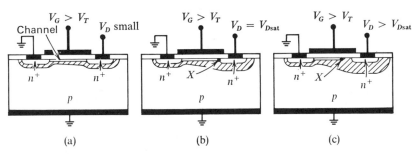

Fig. 11.2 Illustration of the operation of a surface field-effect transistor for $V_G > V_T$.
(a) V_D is small; channel resistance is constant.
(b) $V_D = V_{Dsat}$; onset of saturation.
(c) $V_D > V_{Dsat}$; no further increase in drain current.

characteristics of an n-channel MOS transistor, which we will use for illustration throughout this chapter, shown for various values of the gate voltage in Figure 11.3. The characteristics indeed start out in a straight-line fashion and begin to bend as the drain voltage is increased.

As the drain voltage V_D is increased still further, the voltage drop across the oxide near the drain is further reduced until eventually it falls below the level required to maintain an inversion layer. The drain voltage at which this happens will be denoted by the symbol V_{Dsat}. *At this drain voltage, the channel near the drain disappears.* The surface will be merely depleted and no longer inverted here, as illustrated in Figure 11.2b.

The potential at the end of the inversion layer, at the point X in Figure 11.2b, will be that value for which the gate voltage V_G can still maintain an inversion layer. By our above definition, this value is V_{Dsat}. *Once the drain voltage exceeds V_{Dsat}, the potential at the end of the inversion layer, at the point X, will remain constant, independent of any further increase in the drain voltage,* although the point X will move somewhat toward the source as illustrated in Figure 11.2c.

The current now is due to the carriers that flow down the inversion layer and are injected into the depletion region near the drain. The magnitude of this current will not change significantly with increasing drain voltage since it depends on the potential drop from the beginning of the inversion layer to the end of the inversion layer (point X) and this potential drop remains unaltered.† *Thus, for drain voltages larger than V_{Dsat} the current will not change substantially and will remain at the value I_{Dsat}*, as is evident from the experimental data shown in Figure 11.3.

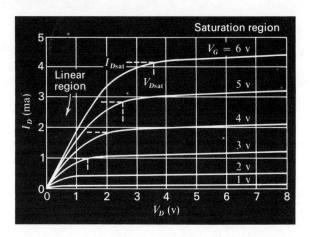

Fig. 11.3 Current-voltage characteristics of a silicon n-channel MOS transistor. This device is used for illustration throughout this chapter. Its structural parameters are: $Z/L = 25$, $x_o = 0.08\ \mu$, $N_A = 3 \times 10^{16}$ cm^{-3}, $V_{FB} = -2.0$ v.

If the gate voltage is increased, the conductance for small values of drain voltage will be larger and the drain voltage at which the current saturates V_{Dsat} will also be larger. As a result, the saturated current will also have a larger magnitude, as is evident in Figure 11.3.

Thus we can distinguish two regions of operation of the surface field-effect transistor. At low drain voltages, the current-voltage characteristics are nearly ohmic or linear (*linear region*), while at high drain voltages the current saturates with increasing drain voltage (*saturation region*). These two regions, the shape of the characteristics, and the manner in which saturation sets in are all reminiscent of junction field-effect transistors.

It is instructive to compare the most important operating principles of the junction and surface field-effect transistors.

† This assumes that the movement of the point X toward the source is negligible. The validity of this assumption will be discussed in a later section.

| JUNCTION FET | SURFACE FET |

- The gate terminal is electrically isolated from channel by:

  reverse-biased depletion region.   insulator.

- The magnitude of the conducting charge is modulated by:

  the width of a reverse biased depletion region.   the incident surface field.

- For small drain voltages, the channel is essentially ohmic. Increasing drain voltage reduces the average magnitude of the conducting charge, thereby reducing the channel conductance.

- When drain voltage exceeds a certain value, the potential drop from source to end of the channel remains at the fixed value $V_{D\text{sat}}$. Hence, the current flow also remains at a fixed value $I_{D\text{sat}}$ for drain voltages $V_D > V_{D\text{sat}}$.

- $V_{D\text{sat}}$ and, therefore, $I_{D\text{sat}}$ depend on the gate voltage applied.

Thus, the two devices are similar in all respects except in the physical mechanism responsible for varying the magnitude of the conducting charge.

11.2 CHARACTERISTICS OF SURFACE FIELD-EFFECT TRANSISTORS

a. Current-Voltage Relationship[5]

Let us consider a surface field-effect transistor operating in the linear region, as shown in Figure 11.4. The voltage drop across an elemental section of the channel is given by

$$dV = I_D\, dR = -\frac{I_D\, dy}{Z\mu_n Q_n(y)} \tag{11.2}$$

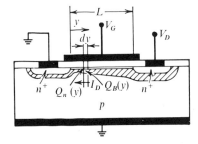

Fig. 11.4 The elemental section of the channel employed in the derivation of the current-voltage characteristics of surface field-effect transistors.

in analogy to Equation 11.1 except that L is replaced by dy. (Note that this equation is identical to the corresponding equation for the junction field-effect transistor except for $-Q_n(y)$ replacing $qN_D[d - 2W(y)]$. However, both quantities represent the charge density of electrons per unit surface area.)

At a distance y from the source, the total charge induced in the silicon Q_s will partly consist of charge in the inversion layer Q_n and partly of charge in the surface depletion region due to the ionized acceptor ions, Q_B. Thus,

$$Q_s(y) = Q_n(y) + Q_B(y).\dagger \qquad (11.3)$$

From Equation 9.16 we have

$$V_G - V_{FB} = -\frac{Q_s}{C_o} + \phi_s \qquad (11.4)$$

which relates the voltage applied to the field plate of an MOS structure to the charge induced in the silicon Q_s and to the surface potential ϕ_s which denotes the total bending of the energy bands. We also include here the flat band voltage V_{FB} which is due partly to the presence of charges in the insulator or at the interface, and partly to a finite metal-semiconductor work function difference as discussed in Chapter 9. $C_o = K_o \epsilon_0 / x_o$ is the capacitance of the oxide layer per unit area.

Combining Equations 11.3 and 11.4 yields the charge density in the inversion layer,

$$Q_n(y) = -[V_G - V_{FB} - \phi_s(y)]C_o - Q_B(y). \qquad (11.5)$$

Since we have assumed that a conducting inversion layer exists, the surface potential ϕ_s will be given approximately by the condition of strong inversion in the presence of an applied voltage. From Equation 10.1, this is

$$\phi_s(y) = V(y) + 2\phi_{Fp} \qquad (11.6)$$

where ϕ_{Fp} is the Fermi potential of the substrate, and $V(y)$ is the reverse bias between the elemental section of the channel and the substrate. The charge within the surface depletion region Q_B is given by

$$Q_B(y) = -qN_A x_{dmax}(y) = -\sqrt{2K_s \epsilon_0 q N_A [V(y) + 2\phi_{Fp}]}, \qquad (11.7)$$

from Equations 10.7 and 10.4.

Since the voltage $V(y)$ increases from the source toward the drain due to the IR drop along the channel, the field-induced junction between the n-type

† In the derivation of the current-voltage characteristics of surface field-effect transistors some authors either neglect $Q_B(y)$ in Equation 11.3 or treat it as a constant. This approximation can lead to significant error in practical devices.

Characteristics of Surface Field-Effect Transistors

inversion layer and the substrate becomes increasingly reverse biased as we proceed from source to drain. Hence, both the energy band bending ϕ_s and the charge within the surface depletion region Q_B increase from source to drain.

Combining Equations 11.2, 11.5, 11.6, and 11.7 and integrating between the source, where $y = 0$† and $V = 0$, and the drain, where $y = L$ and $V = V_D$, yields

$$I_D = \frac{Z}{L}\mu_n C_o \left\{ \left[V_G - V_{FB} - 2\phi_{Fp} - \frac{V_D}{2} \right] V_D \right. $$
$$\left. - \frac{2}{3}\frac{\sqrt{2K_s\epsilon_0 q N_A}}{C_o}[(V_D + 2\phi_{Fp})^{3/2} - (2\phi_{Fp})^{3/2}] \right\} \ddagger \quad (11.8)$$

which can be rewritten in the form

$$I_D = \frac{Z}{L}\mu_n C_o \left\{ \left[V_G - V_{FB} - 2\phi_{Fp} - \frac{V_D}{2} \right] V_D \right. $$
$$\left. - \frac{4}{3}\frac{K_s x_o}{K_o x_{d\max,o}}\sqrt{2\phi_{Fp}}[(V_D + 2\phi_{Fp})^{3/2} - (2\phi_{Fp})^{3/2}] \right\} \quad (11.9)$$

where

$$x_{d\max,o} = \sqrt{\frac{2K_s\epsilon_0(2\phi_{Fp})}{qN_A}}$$

is the width of the depletion region of a field-induced junction *in equilibrium*.

The current-voltage characteristics of the *n*-channel MOS transistor are shown in Figure 11.5 for various gate voltages applied. The corresponding calculated characteristics are also shown in this figure, based on Equation 11.8. This equation is valid only below saturation. Thus the curves shown were calculated for $0 \leqslant V_D \leqslant V_{D\text{sat}}$. Beyond $V_{D\text{sat}}$ the current was taken to be constant, as discussed in Section 11.1.

It is interesting to consider the two limiting forms of Equation 11.8. For very small drain voltages, i.e., $V_D \ll 2\phi_{Fp}$, an expansion of the bracketed terms leads to the formulas,

$$I_D \cong \frac{Z}{L}\mu_n C_o \left[V_G - V_{FB} - 2\phi_{Fp} + \frac{Q_{B,o}}{C_o} \right] V_D \quad \text{[Linear region]} \quad (11.10)$$

---

† We define the potential at the source as our ground potential. In this treatment we only consider the case in which no bias is applied to the substrate relative to the source.
‡ This and other important formulas are summarized at the end of this chapter in Table 11.1 for both *n*- and *p*-channel devices.

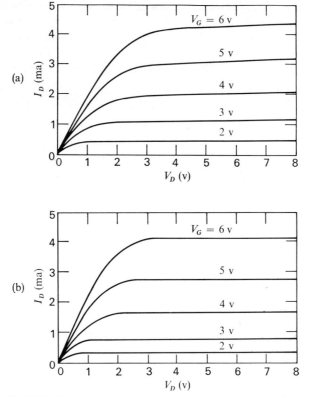

Fig. 11.5 Current-voltage characteristics of the n-channel MOS transistor.
(a) Experiment (same as Fig. 11.3).
(b) Theory (Equation 11.8).

where $Q_{B,o} = -\sqrt{2K_s\epsilon_0 q N_A (2\phi_{Fp})}$ is the charge density per unit area within the surface depletion region, *in equilibrium*. Thus an ohmic characteristic results. The channel conductance

$$g = \frac{\partial I_D}{\partial V_D}\bigg|_{V_G = \text{const}} \quad (11.11)$$

in the linear region is

$$g = \frac{Z}{L}\mu_n C_o (V_G - V_T) \quad \text{[Linear region]} \quad (11.12)$$

where the *turn-on voltage* V_T, i.e., the voltage that must be applied to the gate in order to induce a conducting channel, is given by

$$V_T = V_{FB} + 2\phi_{Fp} - \frac{Q_{B,o}}{C_o}. \quad (11.13)$$

Characteristics of Surface Field-Effect Transistors

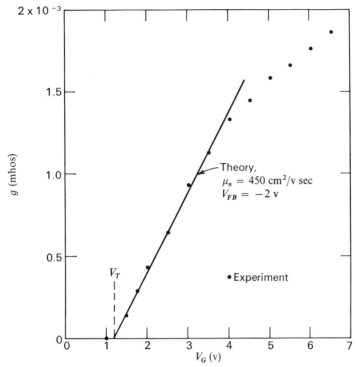

Fig. 11.6 Channel conductance of the MOS transistor as a function of gate voltage, in the linear region (V_D small).

The channel conductance of the MOS transistor in the linear region is shown in Figure 11.6 as a function of gate voltage. The reduction in slope at high gate voltages is due to a decrease in the mobility of electrons owing to increased surface scattering (see Chapter 12).

The above derivation loses its validity when the inversion layer disappears near the drain, i.e., when $V_D = V_{Dsat}$. The condition $Q_n(L) = 0$, inserted into Equation 11.5, yields

$$V_{Dsat} + \frac{1}{C_o}\sqrt{2K_s\epsilon_0 qN_A[V_{Dsat} + 2\phi_{Fp}]} + 2\phi_{Fp} - V_G + V_{FB} = 0 \tag{11.14}$$

where we have also used Equations 11.6 and 11.7, evaluated at $V = V_{Dsat}$.
Solving Equation 11.14 for V_{Dsat} leads to

$$V_{Dsat} = V_G - V_{FB} - 2\phi_{Fp} + \frac{K_s\epsilon_0 qN_A}{C_o^2}\left[1 - \sqrt{1 + \frac{2C_o^2(V_G - V_{FB})}{K_s\epsilon_0 qN_A}}\right] \tag{11.15}$$

or

$$V_{Dsat} = V_G - V_{FB} - 2\phi_{Fp}$$
$$+ 4\phi_{Fp}\left(\frac{K_s x_o}{K_o x_{dmax,o}}\right)^2 \left[1 - \sqrt{1 + \left(\frac{K_o x_{dmax,o}}{K_s x_o}\right)^2 \frac{V_G - V_{FB}}{2\phi_{Fp}}}\right]. \quad (11.16)$$

Note that when the oxide thickness x_o is small in comparison to the width of the surface depletion region $x_{dmax,o}$, this expression reduces to the simple form,

$$V_{Dsat} \cong V_G - V_{FB} - 2\phi_{Fp} \qquad [x_o \ll x_{dmax,o}]. \quad (11.17)$$

Substitution of V_{Dsat} into the current-voltage relationship, Equation 11.8 or 11.9, gives the magnitude of the saturation current, I_{Dsat}.

b. Transconductance

As for junction field-effect transistors, the transconductance is defined by

$$g_m \equiv \frac{\partial I_D}{\partial V_G}\bigg|_{V_D=\text{const}} \quad (11.18)$$

Differentiating Equation 11.9 leads to

$$g_m = \frac{Z}{L}\mu_n C_o V_D \quad (11.19)$$

for $V_D \leqslant V_{Dsat}$.

The transconductance in the saturation range can be calculated by setting $V_D = V_{Dsat}$. This leads to

$$g_{msat} = \frac{Z}{L}\mu_n C_o \left\{ V_G - V_{FB} - 2\phi_{Fp} \right.$$
$$\left. + 4\phi_{Fp}\left(\frac{K_s x_o}{K_o x_{dmax,o}}\right)^2 \left[1 - \sqrt{1 + \left(\frac{K_o x_{max,o}}{K_s x_o}\right)^2 \frac{V_G - V_{FB}}{2\phi_{Fp}}}\right] \right\}$$

[Saturation region]. (11.20)

The transconductance of the MOS transistor in the saturation region is shown as a function of gate voltage in Figure 11.7, in comparison to calculations based on Equation 11.20. As in the case of the conductance in the linear region, a decrease is observed at large gate voltages due to a reduction in surface mobility.

Modification of the Simple Theory

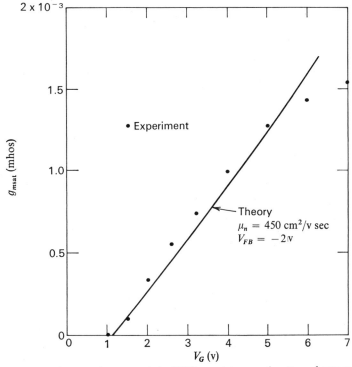

Fig. 11.7 Transconductance of the MOS transistor as a function of gate voltage, in the saturation region ($V_D > V_{Dsat}$).

c. Gate Leakage Current

In the surface field-effect transistor, the gate is insulated from the channel region by a silicon dioxide layer. Because a thermally grown silicon dioxide layer is an excellent insulator, the leakage current between gate and channel region is extremely small, less than 10^{-14} amp in a typical case. Thus the surface field-effect transistor features an exceedingly high input resistance, several orders of magnitude higher even than junction field-effect transistors.

11.3 MODIFICATION OF THE SIMPLE THEORY

a. Effect of Fast Surface States

We have already seen in Chapter 9 that metal-semiconductor work function difference or charges within the oxide or at the interface introduce a displacement of MOS transistor characteristics along the voltage axis by an amount V_{FB}. If, in addition, there are fast surface states present, the charge in these surface states will also result in an additional displacement along the voltage axis. This contribution to the displacement, however, varies with the surface potential (and hence with gate voltage), because the probability of the occupancy of a given surface state itself varies with the surface potential. Thus the displacement of the characteristics along the voltage axis due to fast surface states will not be parallel.

More important, however, is the following: When the carrier concentration near the surface is changed by changing the gate voltage, some of the additional carriers induced near the surface will enter into surface states and therefore will not contribute to an increase in channel conductance. As a result, the observed transconductance will be smaller than the theoretical transconductance. This will be so at frequencies low enough that the surface states can be charged and discharged in phase with the measurement signal. At higher frequencies, when the surface states cannot respond rapidly enough, the observed transconductance will approach the transconductance without surface states.

For this effect to become important, the charge variation in surface states must be comparable to the magnitude of the conducting charge in the channel which is generally of the order of 10^{11} cm^{-2}. Because in thermally oxidized silicon surfaces the surface state density can be reduced to 10^9 to 10^{10} cm^{-2} (see Chapter 12), fast surface states do not pose a significant problem in device operation. This has been a very important factor in making the construction and operation of MOS transistors feasible.

b. Cut-Off Frequency for Transconductance

By exactly the same argument as we have used in Chapter 8 for the junction field-effect transistor, the maximum frequency of operation of a surface field-effect transistor will be given by

$$f_o = \frac{g_m}{C_G} \dagger \qquad (11.21)$$

where C_G is the *total* gate capacitance.

Using Equation 11.19 and $C_G = C_o ZL$ leads to

$$f_o = \frac{\mu_n V_D}{L^2} \qquad (11.22)$$

† More rigorous considerations yield $g_m/2\pi C_G$ for this frequency limitation.

for $V_D \leqslant V_{D\text{sat}}$. Note that a high mobility is desirable for a large cut-off frequency as in the case of junction field-effect transistors.

c. Source-to-Drain Resistance in Saturation[5]

We have seen in the previous section that after saturation the potential at the end of the inversion layer—at the point X—will be fixed at the value $V_{D\text{sat}}$. As the drain voltage is increased further, the reverse bias across the drain junction will increase. As a result, the depletion region separating point X from the drain will widen and the point X will move toward the source, as indicated in Figure 11.2c. Thus the effective channel length will become shorter and, as a result, the drain current will increase slightly with increasing drain voltage, resulting in an upward tilting of the current-voltage characteristics beyond saturation. This phenomenon is particularly prominent for devices with small channel length L.

d. Effect of Series Resistance

The same considerations that were applied in the case of junction field-effect transistors lead to the conclusion that the effect of unmodulated series resistances near the source and the drain will be to lower the observed channel conductance and transconductance as compared to their values without series resistances according to the formulas

$$g(\text{obs}) = \frac{g}{1 + (R_s + R_d)g} \qquad \text{[Linear region]} \qquad (11.23)$$

and

$$g_{m\text{sat}}(\text{obs}) = \frac{g_{m\text{sat}}}{1 + R_s g_{m\text{sat}}}, \qquad \text{[Saturation region]} \qquad (11.24)$$

where g and $g_{m\text{sat}}$ denote the channel conductance and transconductance in the absence of series resistances.

Because in a surface field-effect transistor of the type illustrated in Figure 11.1 all of the channel region is under the influence of the gate, the series-resistance effect is considerably less important than in junction field-effect transistors.

11.4 OTHER TYPES OF SURFACE FIELD-EFFECT TRANSISTORS

The preceding discussion so far was for an n-channel device of the type where we have to apply a positive voltage to the gate in order to bring about the onset of channel conduction. By complete analogy, the discussion could have been applied to a *p*-channel device where a negative gate voltage would have been required to bring about channel conduction. Because with zero gate voltage applied neither of these devices conducts, such devices are called *normally "off"* MOS transistors.

By contrast, a *normally "on"* device can be made by suitable control of the flat-band voltage. Thus, if V_{FB} is sufficiently negative in the case of an *n*-channel device or positive in the case of a *p*-channel device, a conducting channel exists even with zero gate voltage applied. In such a case, the channel conductance can be both increased and decreased with suitable variation of the gate voltage.

The normally "off" *n*- and *p*-channel devices as well as a normally "on" *n*-channel device are illustrated in Fig. 11.8. In all three devices the conducting channel is induced by surface fields, whether the surface fields are due to a gate voltage or to charges within the oxide. Thus we may call these devices *field-induced channel* devices. It is also possible to build-in a channel metallurgically. Two examples of such *metallurgical channel* devices are shown in Figure 11.9. The first of these, illustrated in Figure 11.9a, is made by epitaxially growing an *n*-type film on a *p*-type substrate.[6] The conductivity between the source and drain regions is then modulated

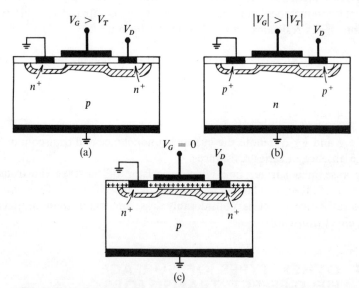

Fig. 11.8 Surface field-effect transistors with field-induced channels.
(a) Normally "off" *n*-channel device.
(b) Normally "off" *p*-channel device.
(c) Normally "on" *n*-channel device.

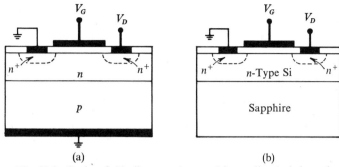

Fig. 11.9 Surface field-effect transistors with metallurgical channels.
(a) Epitaxially grown n-channel device.
(b) Silicon-on-sapphire n-channel device.

by the gate voltage. A positive gate voltage attracts more electrons to the channel, while a negative gate voltage depletes the surface and thereby modulates the cross-sectional area available for current flow. In this mode of operation, this surface field-effect transistor works very much like a junction field effect transistor. A similar device made by growing n-type silicon on an insulating sapphire substrate[7] is illustrated in Figure 11.9b.

READING REFERENCES

MOS transistors are reviewed in *Field Effect Transistors, Physics, Technology and Applications*, J. T. Wallmark and H. Johnson, Eds., Prentice-Hall, 1966.

REFERENCES CITED

1. J. E. Lilienfeld, U.S. Patent 1,745,175 (1930).
2. O. Heil, British Patent 439,457 (1935).
3. W. Shockley and G. L. Pearson, "Modulation of Conductance of Thin Films of Semiconductors by Surface Charges," *Phys. Rev.*, **74**, 232 (1948).
4. D. Kahng and M. M. Atalla, "Silicon-Silicon Dioxide Field Induced Surface Devices," IRE Solid-State Device Research Conference, Carnegie Inst. of Tech., Pittsburgh, 1960.
5. This treatment follows that given by H. K. J. Ihantola, "Design Theory of a Surface Field-Effect Transistor," Stanford Electronics Laboratories Technical Report No. 1661-1 (1961), and by H. K. J. Ihantola and J. L. Moll, "Design Theory of a Surface Field-Effect Transistor," *Solid-State Electronics*, **7**, 423 (1964).

 The characteristics of such devices were also studied by S. R. Hofstein and F. P. Heiman, "The Silicon Insulated-Gate Field-Effect Transistor," *Proc. IEEE*, **51**, 1190 (1963); and by C. T. Sah, "Characteristics of the Metal-Oxide-Semiconductor

Transistors," *IEEE Trans. Electron Devices,* **ED-11,** 324 (1964). For a discussion of a somewhat different type of surface field-effect transistor structure, see, for instance, H. Borkan and P. K. Weimer, "An Analysis of the Characteristics of Insulated-Gate Thin-Film Transistors," *RCA Rev.,* **24,** 153 (1963).

6. V. G. K. Reddi, "Tunable High-Pass Filter Characteristics of a Special MOS Transistor, *IEEE Trans. Electron Devices,* **ED-12,** 581 (1965).

7. C. W. Mueller and P. H. Robinson, "Grown-Film Silicon Transistors on Sapphire," *Proc. IEEE,* **52,** 1487 (1964); F. P. Heiman, "Thin-Film Silicon-on-Sapphire Deep-Depletion MOS Transistors," and S. R. Hofstein, "An Analysis of Deep-Depletion Thin-Film MOS Transistors," *IEEE Trans. Electron Devices,* **ED-13,** (December 1966).

PROBLEMS

11.1 Rederive the current-voltage characteristics, neglecting the charge Q_B within the surface depletion region. Compare with Equation 11.8. Under what condition is this a reasonable approximation? Also, derive the conductance in the linear region and the transconductance in saturation, neglecting Q_B.

11.2 Derive an expression giving the electric field along the channel. Compare its magnitude for the device used for illustration in this chapter with the magnitude of the surface field under various operating conditions.

11.3 The current-voltage characteristics of surface field-effect transistors are often examined in a two-terminal mode, with the drain and gate connected together, and the source and substrate grounded. Derive the current-voltage characteristics in this configuration. Under what conditions is it possible to infer the turn-on voltage V_T from these characteristics?

11.4 A surface field-effect transistor can be operated as a four-terminal device. For example, the source may be reverse biased with respect to the substrate.
(a) Derive the current-voltage characteristics, the conductance, and transconductance in such a case two ways: with substrate held at ground potential, and with source held at ground potential.
(b) What is the turn-on voltage (with respect to ground) in these cases? Compare with results obtained in Chapter 10.
(c) Derive the transductance with respect to the source-substrate voltage.

11.5 In many applications it is important to have a device with a *square-law* characteristic, i.e., $I \propto V^2$. Examine the suitability of the surface field-effect transistor from this standpoint. In particular, consider the best mode of operation and the optimum choice of material and structural parameters for this purpose.

11.6 Derive the dependence of the channel conductance in the linear region on temperature, at a given gate voltage. Assume that the flat-band voltage is independent of temperature, and that the inversion layer mobility is inversely proportional to the absolute temperature.

11.7 Derive an expression for the channel conductance in the linear region in the presence of surface states uniformly distributed in energy throughout the forbidden gap, with density D_{st} (cm^{-2} ev^{-1}). Compare the effect of such surface states with the effect of an inversion layer mobility which is a function of the surface field.

TABLE 11.1
IMPORTANT FORMULAS FOR FIELD-INDUCED CHANNEL SURFACE FIELD-EFFECT TRANSISTORS

	n-Channel $V_D > 0$, $V_G > V_T$, I_D flows from drain to source	p-Channel $V_D < 0$, $V_G < V_T$, I_D flows from source to drain
Current-voltage characteristics	$I_D = \dfrac{Z}{L}\mu_n C_o \left\{\left[V_G - V_{FB} - 2\phi_{Fp} - \dfrac{V_D}{2}\right]V_D \right.$ $\left. - \dfrac{2}{3}\dfrac{\sqrt{2K_s\epsilon_0 q N_A}}{C_o}[(V_D + 2\phi_{Fp})^{3/2} - (2\phi_{Fp})^{3/2}]\right\}$	$I_D = \dfrac{Z}{L}\mu_p C_o \left\{\left[V_G - V_{FB} - 2\phi_{Fn} - \dfrac{V_D}{2}\right]V_D \right.$ $\left. - \dfrac{2}{3}\dfrac{\sqrt{2K_s\epsilon_0 q N_D}}{C_o}[|V_D + 2\phi_{Fn}|^{3/2} - |2\phi_{Fn}|^{3/2}]\right\}$
Saturation voltage	$V_{Dsat} = V_G - V_{FB} - 2\phi_{Fp}$ $+ \dfrac{K_s\epsilon_0 q N_A}{C_o^2}\left[1 - \sqrt{1 + \dfrac{2C_o^2(V_G - V_{FB})}{K_s\epsilon_0 q N_A}}\right]$	$V_{Dsat} = V_G - V_{FB} - 2\phi_{Fn}$ $- \dfrac{K_s\epsilon_0 q N_D}{C_o^2}\left[1 - \sqrt{1 - \dfrac{2C_o^2(V_G - V_{FB})}{K_s\epsilon_0 q N_D}}\right]$
Turn-on voltage	$V_T = V_{FB} + 2\phi_{Fp} + \dfrac{\sqrt{2K_s\epsilon_0 q N_A(2\phi_{Fp})}}{C_o}$	$V_T = V_{FB} + 2\phi_{Fn} - \dfrac{\sqrt{2K_s\epsilon_0 q N_D|2\phi_{Fn}|}}{C_o}$
Conductance (linear region)	$g = \dfrac{Z}{L}\mu_n C_o(V_G - V_T)$	$g = -\dfrac{Z}{L}\mu_p C_o(V_G - V_T)$
Transconductance	$g_m = \dfrac{Z}{L}\mu_n C_o V_D$ for $V_D \leq V_{Dsat}$	$g_m = -\dfrac{Z}{L}\mu_p C_o V_D$ for $|V_D| \leq |V_{Dsat}|$
Maximum frequency	$f_o = \dfrac{g_m}{C_G} \leq \dfrac{\mu_n V_D}{L^2}$ for $V_D \leq V_{Dsat}$	$f_o = \dfrac{g_m}{C_G} \leq \dfrac{\mu_p|V_D|}{L^2}$ for $|V_D| \leq |V_{Dsat}|$
Effect of series resistance	Linear region $\quad g(\text{obs}) = \dfrac{g}{1 + (R_s + R_d)g}$ Saturation region $\quad g_{msat}(\text{obs}) = \dfrac{g_{msat}}{1 + R_s g_{msat}}$	

- FAST SURFACE STATES
- SPACE CHARGE WITHIN OXIDE
- SURFACE-STATE CHARGE
- BARRIER ENERGIES
- SURFACE MOBILITY
- CONDUCTION ON OXIDE SURFACES
- OTHER INSULATORS

12

Properties of the Silicon-Silicon Dioxide System

In the previous three chapters we have discussed the theory of semiconductor surfaces, the effect of surfaces on *p-n* junctions, and surface field-effect transistors. Although most of that discussion is of general validity, its usefulness depends on the availability of specific knowledge of a particular interface system.

As a result of extensive studies using MOS capacitors, gate-controlled diodes, and MOS surface field-effect transistors, the silicon-silicon dioxide system has become a very well characterized interface. A detailed understanding of many of its features is still lacking, but our empirical knowledge of and our ability to control the properties of this interface are quite extensive.

In this chapter, we present a summary of the electrical characteristics of the silicon-silicon dioxide system. We begin with a discussion of the various types of charges and states associated with this system. These are summarized in Figure 12.1. They include *fast surface states* located at the oxide-silicon interface. In addition, there may be space charges present within the oxide layer due to *mobile impurity ions*, e.g., sodium ionic contamination, or due to *traps* ionized by irradiation. Finally, there is a fixed *surface-state charge* located at the interface between oxide and silicon.

Following this, we discuss measurements of the barrier energies associated with the metal-oxide-silicon system and the information they yield about the band structure of this system. We then discuss the mobility of

Fast Surface States

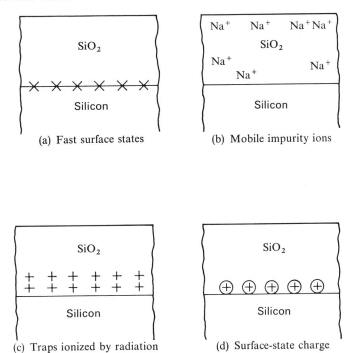

Fig. 12.1 Charges and states associated with the Si/SiO$_2$ system.

electrons and holes in silicon near the interface. The conduction process that may take place on the outer surface of the oxide is considered next. Finally, we briefly review what is known about other insulators used in MIS structures on silicon.

12.1 FAST SURFACE STATES

It has been predicted theoretically, by Tamm and by Shockley,[1] that because of disruption of the periodicity of the lattice at a surface, a high density of states will be introduced into the forbidden gap near a semiconductor surface. Such states have traditionally been called *fast surface states*. Theoretically, it is believed that there should be of the order of one fast surface state for every surface atom, resulting in a density of about 10^{15} cm^{-2}. In fact, experimental observations with surfaces obtained by cleaving under high vacuum bear out this prediction. In contrast, etched germanium and silicon surfaces, which are always covered by a very thin oxide layer, show fast surface-state densities of the order of

10^{11} to 10^{12} cm^{-2}.[1] Initial investigations of thermally oxidized silicon surfaces[2] have found such surfaces to contain fast surface states in densities of the same order.

We saw in Chapter 9 that fast surface states result in a deviation in the *shape* of the capacitance-voltage or channel conductance-voltage characteristics of metal-insulator-semiconductor structures from the ideal theoretical curve shapes. In fact, the experimental characteristics of aluminum-silicon dioxide-silicon structures fabricated by thermal oxidation, including a low-temperature annealing step, follow the theoretical curve shape very closely[3]—although they are displaced from it by a constant amount.

The degree of parallelism of experimental and theoretical characteristics indicates that the total density of surface states in the middle 0.7-ev portion of the silicon energy band is less than 5×10^{10} cm^{-2}. Measurements by different methods—using the a-c conductance of MOS capacitors,[4] and the variation of the turn-on voltage of MOS transistors with temperature[5]—yield results consistent with this upper bound.

In Chapter 10 we saw that the reverse current versus gate voltage characteristic of gate-controlled diodes directly yields the surface recombination velocity s_o, which in turn is related to the fast surface-state density N_{st} by the relation (see also Chapter 5),

$$s_o = \sigma v_{th} N_{st}. \qquad (12.1)$$

Such measurements have yielded values of s_o between 1 and 10 cm/sec.[6a]

Comparison of values of s_o with the corresponding values of the fast surface-state density N_{st} on structures where they were both increased by irradiation[6b] verified Equation 12.1, and yielded values for the capture cross section, σ, of 10^{-16} to 10^{-15} cm^2. With such capture cross sections, the preirradiation values of s_o correspond to fast surface-state densities of 10^9–10^{10} cm^{-2}, which are consistent with the upper bound obtained by the other methods.

The above values of fast surface-state density were all obtained on aluminum-thermally grown silicon dioxide-silicon structures which were subjected to a heat treatment step at about 500°C. Such a procedure has been shown[7] to lead to an order-of-magnitude reduction in the density of fast surface states.

For simplicity, this discussion has been in terms of single-level surface states. Actually, it appears that the fast surface states are more or less uniformly distributed in energy over the center portion of the energy gap.[4] As we have seen in Chapter 10, in such a case only those centers which are within a few kT in energy of the middle of the energy gap

Space Charge within the Oxide

contribute effectively to surface recombination and generation. Calculations show[6a] that N_{st} in Equation 12.1 is then replaced by $\pi k T D_{st}$ where D_{st} is the density (cm^{-2} ev^{-1}) of the uniformly distributed centers.

In the case of samples which were not annealed, the existence of large densities of single-level surface states slightly below the conduction band edge and slightly above the valence band edge have also been observed.[8] Because of the location of these states, they do not significantly affect the characteristics of semiconductor devices. (In fact, for the same reason a special technique had to be developed for their measurement.[8])

The relatively low density of fast surface states at thermally oxidized silicon surfaces has been an important advantage of the planar technology.

12.2 SPACE CHARGE WITHIN THE OXIDE

a. Ionic Contamination

A major difficulty encountered with early MOS devices was that the flat-band voltage was unstable, i.e., it was subject to *drift* under bias at elevated temperatures. An example of this drift behavior is illustrated in Figure 12.2. Here the capacitance-voltage characteristics observed initially are marked by (1), while those observed after 30 minutes at 127°C, with $V_G = +10$ v applied, are marked by (2). The characteristics could be recovered; those observed after partial recovery obtained by heating for 30 minutes at the same temperature with the gate shorted to the substrate are marked by (3).

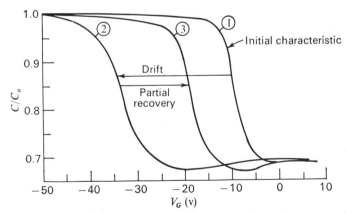

Fig. 12.2 Shift in the capacitance-voltage characteristics due to ionic contamination.[9]

This drift process was shown to be due to the *rearrangement* of an ionic space charge distribution within the oxide.[9] We have seen in Chapter 9 that a contribution to the flat-band voltage,

$$\frac{1}{C_o} \int_0^{x_0} \frac{x}{x_0} \rho(x)\, dx,$$

results when a space charge of distribution $\rho(x)$ is present in the oxide. Thus the drift of the capacitance-voltage characteristics can be due simply

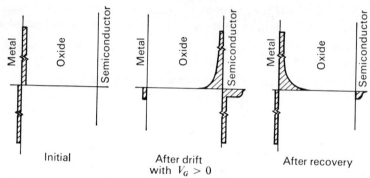

Fig. 12.3 Charge distributions pictured to correspond to various stages of the drift process.[9]

to the rearrangement of the distribution $\rho(x)$ within the oxide without any change in the total density of ions per unit area,

$$Q_o \equiv \int_0^{x_0} \rho(x)\, dx$$

within the oxide. This rearrangement is illustrated in Figure 12.3 where we show the initial charge distribution (where all of the positive ionic space charge is pictured to be next to the metal, exerting no influence on the silicon), the distribution after positive bias drift (resulting in all of the oxide space charge being located next to the silicon and therefore inducing its image charge in the silicon), and finally the distribution obtained after the recovery of the ionic space charge back to the vicinity of the metal electrode.

This picture of a rearrangement of an ionic charge within the oxide is in agreement with all of the experimental observations. It fits the fact that the maximum change in flat-band voltage is independent of the temperature or the magnitude of the positive bias applied during the experiment even though the *rate* of the process itself is dependent on both. It is in agreement with the observation that the time-integral of the current

Space Charge within the Oxide

flowing to the gate during drift is identical with the charge given by $\Delta V_{FB} C_o$. Finally, it is in agreement with the results of experiments in which part of the oxide was gradually removed after drifting. Such experiments showed that after drifting all of the positive charge was located next to the silicon, as shown in the middle portion of Figure 12.3.

It was inferred[9] both from the rate of the drift process and from experiments in which the oxide layer was intentionally contaminated with sodium chloride that the uncontrolled drift in MOS structures was due to trace contamination by sodium. Subsequently, this has been verified by radioactive tracer measurements.[10] Such measurements have also shown that

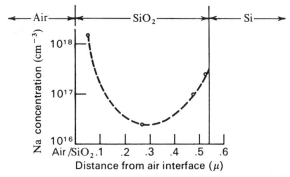

Fig. 12.4 Sodium concentration distribution in the oxide after drift.[10]

the distribution of the sodium ions in the oxide at various stages of the drift process indeed corresponds to the postulated distributions shown in Figure 12.3.

An example of the sodium distribution in an oxide layer which was *not* intentionally contaminated with sodium, in an intermediate stage of the drift process, is shown in Figure 12.4. It is interesting to compare this figure with the results of numerical computations[11] dealing with the transient ion transport problem in the oxide layer. The results of such computations are shown in Figure 12.5 for the particular case of +10 volts applied to the gate at 127°C for various lengths of time. Note that the beginning and the final distributions correspond to the first and second sketches shown in Figure 12.3. Between these extremes, it is evident that the sodium moves across the oxide with a U-shaped distribution prevailing in the intermediate stages, in qualitative agreement with the experimental results shown in Figure 12.4.

The mobile sodium impurity, being due to an external contamination, could be eliminated by appropriate precautions in the device fabrication procedure. Once this contamination was eliminated, MOS devices became stable even under bias at elevated temperatures.[3,12]

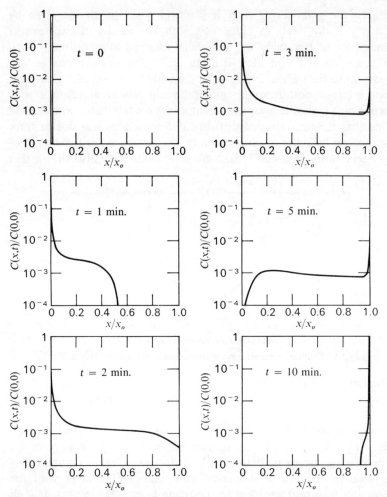

Fig. 12.5 Theoretical sodium concentration distribution in the oxide as a function of time.[11]

b. Radiation-Induced Space Charge

A positive space-charge build-up in silicon dioxide films has been observed to result from exposure to ionizing radiation of various kinds, including X-ray, gamma ray, low- and high-energy electron irradiation.[13] The physical origin of this charge is entirely different from the ionic contamination discussed above.

Experimental results[14] pertaining to X-irradiated silicon dioxide layers are shown in Figure 12.6 as a function of time, for various values of gate

Surface-State Charge

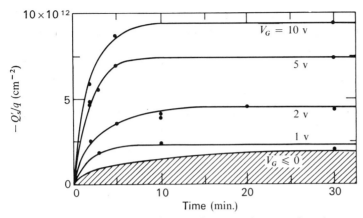

Fig. 12.6 The build-up of excess charge as a function of irradiation time.[14]

voltage applied during irradiation. It is evident that the excess charge Q_s' induced in the silicon by the space charge in the oxide first increases with time and then reaches a steady-state or saturation value which, in turn, is dependent on the gate bias applied during irradiation.

An idealized model for the build-up of such a space charge in the oxide under ionizing radiation exposure is as follows.[14] During irradiation,

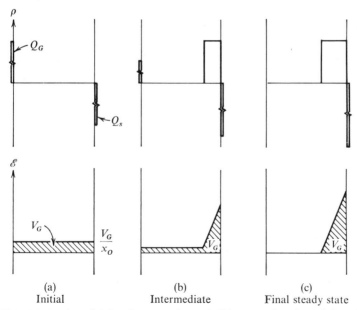

Fig. 12.7 Idealized model for the space-charge build-up as a function of time during irradiation of an MOS structure under a positive gate bias.[14]

electron-hole pairs will be generated in the oxide. If there is no electric field present in the oxide, the electrons and holes will recombine, resulting in no net charge building up in the oxide. However, if there is an electric field present in the oxide, this field will tend to separate the electrons and holes. In particular, a field corresponding to a positive gate voltage will tend to pull electrons toward the gate electrode. If no electrons can enter from the silicon into the oxide, trapping of the holes near the oxide-silicon interface will result in a gradual build-up of a space charge as shown in parts (a) and (b) of Figure 12.7. Due to the increased electric fields, an increasing fraction of the total applied voltage will be dropped across this space-charge region. Thus, as the space charge grows, eventually the field in the rest of the oxide layer is brought to zero. This results in a steady-state situation, depicted in Figure 12.7c.

The radiation-induced oxide space charge can be annealed out relatively rapidly by heating above 300°C.

12.3 SURFACE-STATE CHARGE

Experiments with MOS structures[3] showed the unexpected existence of a fixed charge, apparently located near the interface. This charge results in a parallel translation of the capacitance-voltage (or channel conductance-voltage) characteristics along the voltage axis. It is called the *surface-state charge*, and its density per unit area is designated by the symbol Q_{ss}. The characteristics of the surface-state charge have since been experimentally studied very extensively.[15]

The surface-state charge has the following properties:

1. It is fixed; it cannot be charged or discharged over a wide variation of bending of the silicon energy bands.

2. It is unchanged under conditions that would lead to the motion of sodium ions in the oxide, or to the annealing of radiation-induced space charges.

3. It is located within 200 Å of the oxide-silicon interface.

4. Its density Q_{ss} is not significantly affected by the oxide thickness, or by the type or concentration of impurities in the silicon.

5. Q_{ss} is a strong function of the oxidation and annealing conditions, and the orientation of the silicon crystal.

The orientation effect[16] can be summarized in an approximate fashion as follows: the ratio of surface-state charge density Q_{ss} values under a given oxidation condition for silicon oriented along the (111), (110), and (100) directions will be in the ratio of approximately 3:2:1. This is in

Surface-State Charge

rough agreement with the order of variation of the linear oxidation rate constant B/A for these three orientations (see Chapter 2).

The effect of oxidation and annealing conditions is summarized in Figure 12.8. Here we show the surface-state charge density Q_{ss} resulting after oxidation in dry oxygen or wet oxygen, and after heat treatment of both dry and wet oxides in a nitrogen ambient. The results presented here are extremely reproducible. A further feature of these results is that the only relevant heat treatment is the final one. Regardless of the previous history of a sample, the final heat treatment will determine the value of Q_{ss}, provided only that sufficient time is allowed for the sample to reach steady state under the given conditions.

The above results point to the role of *excess ionic silicon in the oxide* in the origin of the surface-state charge.[15] It appears that the surface-state charge is due to excess ionic silicon present in the oxide during oxidation, waiting to react with the oxidizing species that has diffused across the oxide during the oxidation process, as illustrated in Figure 12.9.

An additional indication of the role of excess silicon in the oxide has been provided by a series of experiments in which the value of Q_{ss} was reproducibly and markedly increased by heating samples under *negative* gate voltage conditions. (Note that this condition is the opposite of that required to bring about a similar effect due to ionic contamination.) The rate of this process is about four orders of magnitude smaller than the rate of sodium rearrangement. Such results are shown in Figure 12.10,

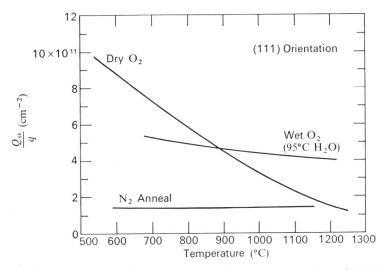

Fig. 12.8 Dependence of the surface-state charge density on the ambient and temperature of the final heat treatment.[15]

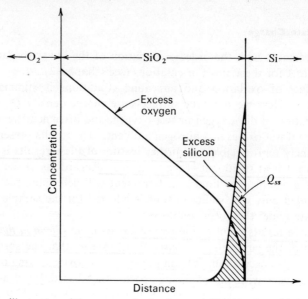

Fig. 12.9 Illustration of the proposed role of excess silicon in the oxide leading to the surface-state charge.[15]

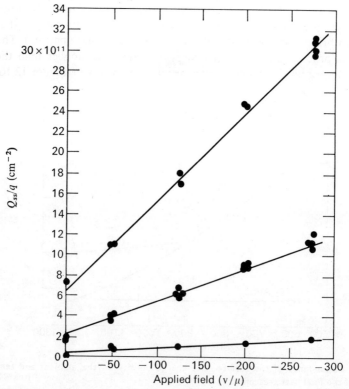

Fig. 12.10 Dependence of the final value of the surface-state charge density on the applied field across the oxide and on the initial surface-state charge density value.[15]

12.4 BARRIER ENERGIES

where we show the final steady-state value of Q_{ss} as a function of the applied field across the oxide. The final surface-state charge density is evidently proportional to both the applied field and the initial surface-state charge density.

12.4 BARRIER ENERGIES

As discussed in Chapter 9, a difference between the metal and semiconductor work functions results in a shift of the flat-band voltage of an MOS structure. This shift provides one method of the evaluation of this work function difference as illustrated in Figure 12.11. In this figure, we

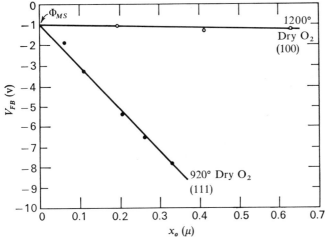

Fig. 12.11 Determination of metal-semiconductor work function difference from MOS flat-band voltage measurements.[15] [Aluminum-SiO$_2$-p type Si ($N_A = 10^{16}$ cm^{-3})]

plot the flat-band voltage V_{FB} as a function of oxide thickness for samples prepared under the same conditions. Since the flat-band voltage is given by

$$V_{FB} = \Phi_{MS} - Q_{ss}\frac{x_o}{K_o\epsilon_0},$$

the fact that a straight line is obtained indicates that the surface-state charge density Q_{ss} is independent of the oxide thickness in agreement with our previous discussion. The intercept of the lines corresponds to the work function difference, which is of the order of 1 volt for the present case of an aluminum-silicon dioxide-p-type silicon ($N_A = 10^{16}$ cm^{-3}) sample.

A method of evaluating the individual barrier energies is based on the measurement of the photoelectric thresholds at the metal-oxide and the oxide-silicon interfaces.[17-19] In this method, the MOS structure is illuminated with light of increasing energy under a given bias condition. When the photon energy becomes sufficiently large to raise electrons from the filled valence band to the conduction band of the oxide, conduction begins across the oxide. The photon energy then corresponds to the barrier energy of the particular interface studied.

Barrier energies determined by both the photoelectric response method and by the MOS method are tabulated in Table 12.1. We also show for

TABLE 12.1

METAL-SiO$_2$ AND Si-SiO$_2$ BARRIER ENERGIES AS MEASURED BY PHOTOELECTRIC EXCITATION AND AS INFERRED FROM MOS FLAT-BAND VOLTAGES. ALSO SHOWN ARE THE CORRESPONDING VACUUM WORK-FUNCTION VALUES.[19]

Metal	Φ_M (Photo)	Φ_M (MOS)	Vacuum work function
Mg	2.25	2.4	3.70
Al	3.2	3.2†	4.20
Ni	3.7	3.6	4.74
Cu	3.8	3.8	4.52
Ag	4.15	4.2	4.31
Au	4.1	4.1	4.70
Si	4.35		5.15

† Arbitrarily chosen as reference.

comparison the vacuum work function values of the respective materials. Note that the barrier energies follow roughly the same order as the work functions. The presence of the oxide merely seems to result in a lowering of the barrier energy as compared to vacuum by 0.5 to 1 ev.

12.5 SURFACE MOBILITY

The mobility of electrons and holes in inversion layers has been studied by using channel conductance measurements on MOS transistor structures.[20] From the slope of the conductance versus gate voltage plots, the mobility of carriers in the inversion layer was evaluated. A typical set of results for both electrons and holes is shown in Figure 12.12. Over a fairly large range of variation of the charge per unit area induced in the silicon Q_s (or equivalently, of the surface field $\mathscr{E}_s$) the inversion-layer mobility is seen to be relatively constant, having a value approximately half that of the

Conduction on Oxide Surfaces

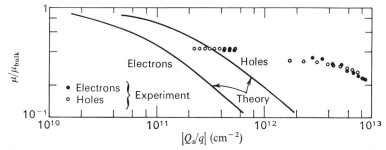

Fig. 12.12 Comparison between typical experimental inversion layer mobilities and theory for diffuse surface scattering.[20]

corresponding bulk mobilities. At higher surface fields, the mobility begins to decrease slightly. These results are in disagreement with the theory of diffuse surface scattering,[21] the predictions of which are also illustrated in Figure 12.12.

The temperature dependence of the inversion-layer mobility in the constant-mobility region is found to be[20] $\mu \propto T^{-1.5}$. This is the temperature dependence predicted for bulk mobility by the theory of lattice scattering (see Chapter 4) which, however, had never been previously observed.

12.6 CONDUCTION ON OXIDE SURFACES

Early investigators[22] of oxidized silicon surfaces have suggested that conduction on the outer surface of the oxide may play an important role in determining the characteristics of underlying silicon devices. This oxide surface conduction process is illustrated schematically in Figure 12.13 where we show a gate-controlled n^+p diode structure with the gate biased positively. Because of conduction on the outer surface of the oxide, its potential will approach that of the metal gate. This is indicated in the upper portion of the figure where we show the potential of the oxide surface for increasing times. In essence, the "gate" will be extended to a larger area. This in turn will result in the formation of a field-induced junction under this "extended gate." (The growth of the field-induced junction can actually be used to study the rate of the surface conduction process.[23])

It can be shown[24] that the surface conduction process is described by an equation of the same *form* as the diffusion equation, with the diffusivity replaced by $1/R_\square C_o$, where $R_\square$ is the sheet resistance of a square-shaped portion of the outer surface of the oxide and C_o is the oxide capacitance.

Accordingly, the oxide surface potential in the case shown in Figure 12.13 will follow an erfc-type distribution. Rate studies of the charging process can thus be used to determine the sheet resistance of the oxide surface $R_\square$.

This picture of the oxide surface conduction has been verified in detail[24] by an experimental technique which gives the potential of the oxide surface directly. Using glass slides and this potential measuring technique, the

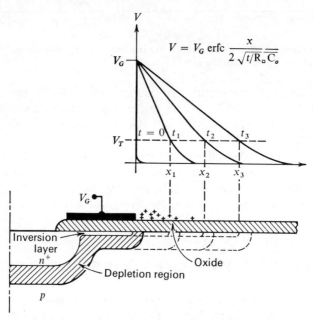

Fig. 12.13 Creation of a field-induced junction by charging outer surface of oxide from a metal gate.[23]

sheet resistance of glass surfaces has been studied[25] as a function of the humidity of the atmosphere, a factor which affects the surface conduction process very drastically. The results are shown in Figure 12.14 along with data obtained on thermally oxidized silicon surfaces by measuring the rate of growth of field-induced junctions.[23] The two sets of data are in reasonably good agreement and show the tremendous influence of humidity in excess of 40% on the sheet resistance. Since the time constant of the oxide surface conduction process is given by $\sim R_\square C_o$, the oxide charging process will proceed very rapidly in a humid atmosphere, and very slowly in a dry atmosphere.

An interesting consequence of the oxide surface conduction process is its effect on MOS characteristics as illustrated in Figure 12.15. The surface conduction process tends to increase the inversion layer area. Thus as this process continues, the silicon capacitance in parallel with the oxide capacitance will grow with time. Correspondingly, the series combination of the two capacitances will approach the capacitance of the oxide layer

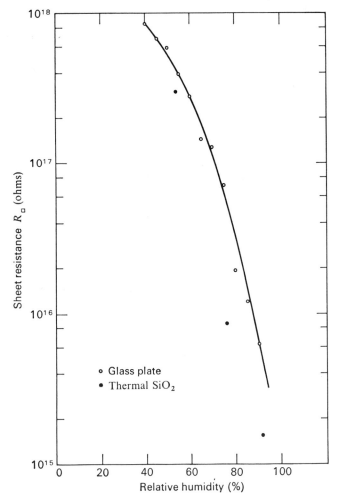

Fig. 12.14 Sheet resistance of glass plate and of thermal SiO$_2$ vs relative humidity.[23,25]

alone because of the increasingly large silicon capacitance up to much higher measurement frequencies than otherwise.[26] Thus, as indicated in Figure 12.15, the shape of the MOS capacitance-voltage characteristics will approach the low-frequency type characteristics.

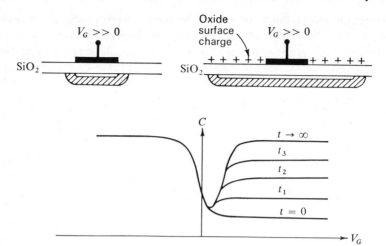

Fig. 12.15 The effect of oxide surface conduction on MOS capacitance-voltage characteristics.

12.7 OTHER INSULATORS

In order to fully utilize the potential capabilities of metal-insulator-semiconductor structures, certain combinations of insulators may be more advantageous than the metal-silicon dioxide-silicon system itself. Of the numerous possibilities, several have already been studied.

It has been found[27] that if a phospho-silicate glass, i.e., SiO_2 rich in P_2O_5, is present on the outside of a silicon dioxide layer, the instability of contaminated MOS devices is substantially reduced. Experiments with

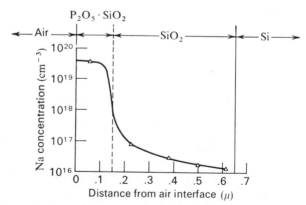

Fig. 12.16 The effect of a P_2O_5-rich layer on the sodium concentration distribution within the oxide.[10]

Other Insulators

radioactive tracers[10] showed that this effect is due to a much higher solubility of sodium in the phospho-silicate glass than in the silicon dioxide. Thus sodium tends to segregate in the phospho-silicate glass layer and, therefore, it is not available for drift across the underlying silicon dioxide layer.

Experimental measurements[10] of the sodium distribution within a phospho-silicate glass/silicon dioxide double layer system are shown in Figure 12.16. Note the three orders of magnitude enhancement of the sodium concentration in the phospho-silicate glass layer relative to the underlying silicon dioxide film.

It has also been found[28] that, independently of sodium contamination, there is a dipole-type polarization process associated with the phospho-silicate layer itself. This dipole-type polarization can be shown to yield a shift in the flat band voltage, the steady-state value of which is given by

$$\Delta V_{FB} = -\alpha \left(\frac{x_g}{x_o}\right) V_P \qquad (12.2)$$

where the proportionality constant α depends on the phospho-silicate glass-to-oxide thickness ratio, x_g/x_o. This equation shows that the shift in the flat-band voltage can take place in either direction, depending on the polarity of the gate voltage applied during the polarization process V_P.

Experimental measurements obtained with a sample which consisted of a thick phospho-silicate glass layer on a thin silicon dioxide layer are shown in Figure 12.17. Note the displacements of the MOS characteristics in either direction from the original one, which is denoted by $V_P = 0$.

Another double-layer dielectric system, which has been studied in detail, is the lead-silicate glass/silicon dioxide double layer.[29] This double-layer system also showed considerable polarization effects which were due to the rearrangement of lead ions. These ions are originally uniformly distributed along with their negative anions in the lead-glass film. If a polarizing voltage is applied, the lead ions can be rearranged, while the anions, which are much less mobile, cannot. Because of this, a space charge will build up in the lead-glass layer, which leads to an induced image charge in the silicon and, correspondingly, to a shift in the flat-band voltage. The steady-state value of this shift is given by

$$\Delta V_{FB} = -\frac{K_o x_g}{K_g x_o} V_P. \qquad (12.3)$$

Thus this type of polarization will also be symmetrical and its direction will depend on the polarity of the applied polarizing voltage.

The drift mechanisms associated with these double-layer dielectrics are illustrated in comparison to the sodium drift mechanism in silicon dioxide in Figure 12.18. In this figure, we show the initial charge distribution in the insulator, and the charge distribution after a negative and a positive polarizing voltage.

Case (a) corresponds to sodium contamination in thermally grown silicon dioxide. Initially, the positive sodium ions are at the outer interface. Their distribution is not changed substantially when a negative voltage is applied to the sample, but the sodium ions can be driven across to the oxide-silicon interface with the application of a positive gate voltage. Case (b) depicts an insulator containing an initially uniform distribution of mobile positive ions compensated by immobile negative ions corresponding to the case of lead glass. This distribution can be disturbed by either negative or positive gate voltage. The positive ions will pile up near the negative terminal in either case, leaving behind a negative space charge near the positive terminal. The combined space charges will then induce an image charge in both the metal and the silicon as indicated in the figure. Case (c) depicts polarization due to orientation of dipoles, an example of which is provided by the phospho-silicate glass. A uniform dipolar polarization can be considered to be equivalent to equal and opposite surface charges appearing on the two faces of the dielectric. These then induce an opposite image charge in the metal and in the semiconductor as illustrated.

Another silicon compound, silicon nitride (Si_3N_4), has been studied both as a possible insulator in MIS structures[31] and in conjunction with silicon dioxide.[32] Silicon nitride has the attractive feature that the diffusivity of various impurities in it—in particular, that of sodium[33]—are

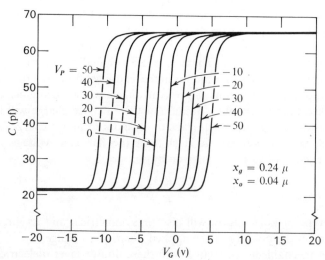

Fig. 12.17 Polarization phenomenon associated with phospho-silicate layers.[28]

References Cited

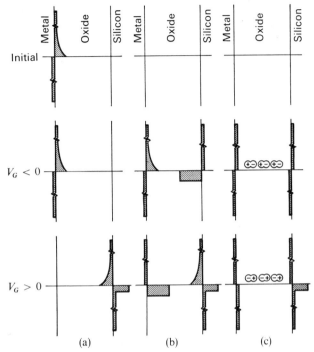

Fig. 12.18 The charge distributions in an MOS structure initially, and after drifting to saturation under negative and positive gate bias. Three cases are shown:
(a) Corresponds to a thermally produced oxide contaminated with mobile positive ions at the outer surface.
(b) Corresponds to a glass containing an initially uniform distribution of mobile positive ions compensated by immobile negative ions.
(c) Corresponds to a glass with a uniform dipolar polarizability. A uniform dipolar polarization is equivalent to equal and opposite surface charges on the two faces of the dielectric.[30]

much lower than in silicon dioxide. Thus surface field-effect transistors made with silicon nitride would be less susceptible to the ionic contamination problem described earlier than those made with silicon dioxide. However, it appears that silicon nitride is inferior to silicon dioxide in other ways: it is much more conductive, and contains large densities of traps which lead to instability of the flat-band voltage of MIS devices.[31,34]

REFERENCES CITED

1. A. Many, Y. Goldstein, and N. B. Grover, *Semiconductor Surfaces*, Wiley, 1965, Chapters 5 and 9.

2. M. M. Atalla, E. Tannenbaum, and E. J. Schreibner, "Stabilization of Silicon Surfaces by Thermally Grown Oxides," *Bell System Tech. J.*, **38**, 749 (1959).

3. A. S. Grove, B. E. Deal, E. H. Snow, and C. T. Sah, "Investigation of Thermally Oxidized Silicon Surfaces Using Metal-Oxide-Semiconductor Structures," *Solid-State Electronics*, **8**, 145 (1965).

4. E. H. Nicollian and A. Goetzberger, "MOS Conductance Technique for Measuring Surface State Parameters," *Appl. Phys. Letters*, **7**, 216 (1965).

5. L. Vadasz and A. S. Grove, "Temperature Dependence of MOS Transistor Characteristics Below Saturation," *IEEE Trans. Electron Devices* **ED-13**, 863 (December, 1966).

6a. A. S. Grove and D. J. Fitzgerald, "Surface Effects on *P-N* Junctions: Characteristics of Surface Space-Charge Regions under Non-Equilibrium Conditions," *Solid-State Electronics*, **9**, 783 (1966).

6b. D. J. Fitzgerald and A. S. Grove, "Radiation-Induced Increase in Surface Recombination Velocity of Thermally Oxidized Silicon Structures," *Proc. IEEE* (Correspondence) **54**, 1601 (1966).

7. G. Cheroff, F. Fang, and F. Hochberg, "Effect of Low Temperature Annealing on the Surface Conductivity of Si in the Si-SiO$_2$-Al System," *IBM Journal*, **8**, 416 (1964); P. Balk, "Effects of Hydrogen Annealing on Silicon Surfaces," Electrochemical Society Spring Meeting, San Francisco, May 1965, Abstract 109.

8. P. V. Gray and D. M. Brown, "Density of SiO$_2$-Si Interface States," *Appl. Phys. Letters*, **8**, 31 (1966).

9. E. H. Snow, A. S. Grove, B. E. Deal, and C. T. Sah, "Ion Transport Phenomena in Insulating Films," *J. Appl. Phys.*, **36**, 1664 (1965).

10. E. Yon, W. H. Ko, and A. B. Kuper, "Sodium Distribution in Thermal Oxide on Silicon by Radiochemical and MOS Analysis," *IEEE Trans. Electron Devices*, **ED-13**, 276 (1966).

11. D. A. Tremere, unpublished.

12. P. Lamond, J. Kelley, and M. Papkoff, "Stable MOS Transistors," *Electro-Technology*, Dec. 1965, p. 40.

13. For a review of "Surface Effects of Radiation on Semiconductor Devices," see J. P. Mitchell and D. K. Wilson, *Bell System Tech. J.*, **46**, 1 (1967).

14. A. S. Grove and E. H. Snow, "A Model for Radiation Damage in Metal-Oxide-Semiconductor Structures," *Proc. IEEE* (Correspondence), **54**, 894 (1966).

15. B. E. Deal, M. Sklar, A. S. Grove, and E. H. Snow, "Characteristics of the Surface-State Charge (Q_{ss}) of Thermally Oxidized Silicon," *J. Electrochemical Soc.*, **114**, 266 (March 1967).

16. P. Balk, P. J. Burkhardt, and L. V. Gregor, "Orientation Dependence of Built-In Surface Charge on Thermally Oxidized Silicon," *Proc. IEEE* (Correspondence), **53**, 2133 (1965).

17. R. Williams, "Photoemission of Electrons from Silicon into Silicon Dioxide," *Phys. Rev.*, **140**, A569 (1965).

18. A. M. Goodman, "Photoemission of Electrons from Silicon and Gold into Silicon Dioxide," *Phys. Rev.*, **144**, 588 (1966).

19. B. E. Deal, E. H. Snow, and C. A. Mead, "Barrier Energies in Metal-Silicon Dioxide-Silicon Structures," *J. Phys. Chem. Solids*, **27,** 1873 (1966).
20. O. Leistiko, A. S. Grove, and C. T. Sah, "Electron and Hole Mobilities in Inversion Layers on Thermally Oxidized Silicon Surfaces," *IEEE Trans. Electron Devices*, **ED-12,** 248 (1965).
21. J. R. Schrieffer, "Effective Carrier Mobility in Surface-Space Charge Layers," *Phys. Rev.*, **97,** 641 (1955).
22. M. M. Atalla, A. R. Bray, and R. Lindner, "Stability of Thermally Oxidized Silicon Junctions in Wet Atmospheres," *Proc. IEE*, **106,** Part B, Supplement No. 17, 1130 (1960).
23. E. H. Snow, in "A Study of Failure Mechanisms in Silicon Planar Transistors," Technical Documentary Report Dec. 1965, RADC Contract AF30 (602)-3776.
24. W. Shockley, W. W. Hooper, H. J. Queisser, and W. Schroen, "Mobile Electric Charges on Insulating Oxides with Application to Oxide Covered *P-N* Junctions," *Surface Science*, **2,** 277 (1964).
25. W. Schroen and W. W. Hooper, in "Failure Mechanisms in Silicon Semiconductors," Final Report RADC AF30 (602)-3016 (1964).
26. E. H. Nicollian and A. Goetzberger, "Lateral AC Current Flow Model for Metal-Insulator-Semiconductor Capacitors," *IEEE Trans. Electron Devices*, **ED-12,** 108 (1965); S. R. Hofstein and G. Warfield, "Physical Limitations on the Frequency Response of a Semiconductor Surface Inversion Layer," *Solid-State Electronics*, **8,** 321 (1965).
27. D. R. Kerr, J. S. Logan, P. J. Burkhardt, and W. A. Pliskin, "Stabilization of SiO_2 Passivation Layers with P_2O_5," *IBM Journal*, **8,** 376 (1964).
28. E. H. Snow and B. E. Deal, "Polarization Phenomena and Other Properties of Phosphosilicate Glass Films on Silicon," *J. Electrochem. Soc.*, **113,** 263 (1966).
29. E. H. Snow and M. E. Dumesnil, "Space Charge Polarization in Glass Films," *J. Appl. Phys.*, **37,** 2123 (1966).
30. B. E. Deal, E. H. Snow, and A. S. Grove, "Properties of the Silicon Dioxide-Silicon System," *SCP and Solid State Technology*, **9,** 25 (1966).
31. S. M. Hu, "Properties of Amorphous Silicon Nitride Films," *J. Electrochem. Soc.*, **113,** 693 (1966).
32. G. H. Schneer, W. vanGelder, V. E. Hauser, and P. F. Schmidt, "A Silicon Nitride Junction Seal on Silicon Planar Transistors," Paper 5.3 at the IEEE Electron Devices Meeting, Washington, October, 1966.
33. J. V. Dalton, "Sodium Drift and Diffusion in Silicon Nitride Films," *J. Electrochem. Soc.*, **113,** 165C (1966).
34. See papers (Abstract Nos. 146-163) given at the Electrochemical Society Fall Meeting, Philadelphia, October, 1966.

Index

Acceptors, 97
Accumulation, 265
Activation energy, epitaxial growth, 12
   intrinsic carrier concentration, 95
   oxidation, 28, 29
   solid-state diffusion, 39
Alloy junction, 2
Amplifier, transistor, 213
Anodization, 22
Atomic or molecular weight, of Ge, Si, GaAs, and SiO_2 (Table), 102
Autodoping, 83
Avalanche breakdown, 191–194
Average resistivity of diffused layers, 54–57

Band theory, 91–95
Band-to-band recombination, 128
Barrier energies in MOS structures, 345–346
Base, factor, 218
   of junction transistors, 210
   resistance, 228–230
Base-width modulation (Early effect), 226
Bias, forward, 150, 161, 180–191
   reverse, 150, 161, 172–180
Bipolar transistor (see Junction transistors)
Boltzmann distribution, 99
Boltzmann statistics, 99, 100
Boundary layer, theory, 14–18
   thickness, 16
"Box" impurity distribution, 54
Breakdown, avalanche, 191–194
   critical field, for Ge, GaAs, Si, and SiO_2 (Table), 103

Breakdown, critical field
   for Si, 193
   of p-n junctions, 150, 191–201
   soft, 200
   Zener (tunneling), 191
Breakdown voltage, common-base, 230
   common-emitter, 230–234
   of diffused junctions, 196
   of field-induced junctions, 305
   of linearly graded junctions, 195
   of one-sided step junctions, 194
   of planar junctions, 197
   reach-through limited, 199
   surface effects on, 311
Bubble analogy, 94
Built-in electric field, 32, 59–63, 224
Built-in voltage, of field-induced junctions, 269, 291
   of linearly graded junctions, 165
   of step junctions, 157

Capacitance, 169
   of linearly graded junctions, 171
   of one-sided step junctions, 171
   small-signal, 169
Capacitance-voltage characteristics, of MOS structures, 271–276
   frequency effects on, 274
   of p-n junctions, 169–172
Capture cross section, 131
   of gold, 141
   of surface states, 145
Carrier concentration, intrinsic, 96, 101
Carrier removal, 142, 144
Centers, recombination-generation, 129
Channel, characteristic, 305

Channel, conductance, junction
field-effect transistors, 245, 250
surface field-effect transistors, 276,
318, 324
currents, 305
junction field-effect transistors, 244
metallurgical, 330
surface field-effect transistors, 318
Charges in insulator, 279–282, 337–341
Chemical potential, of electrons, 98, 156
Chemical reaction, first-order, 10
Chemical surface-reaction rate constant,
10, 25
Chemical transport, 44
Collector of junction transistors, 210
Collisions, time interval between, 107
Common-base mode, 212
breakdown voltage, 230
current gain, 211, 219
Common-emitter mode, 212
breakdown voltage, 230–234
current gain, 211, 220
Complementary error function, 46
Complete ionization, 97
Concentration gradient, at junction, 48,
50, 165
Concentrations, electron, 100
hole, 100
intrinsic carrier, 96, 101
Conduction, 92
band, 92
band edge, 94
electron, 93
on oxide surfaces, 347–350
Conductivity type, 98
Conductivity modulation, 227
Continuum, of recombination-
generation centers, 302
of surface states, 302
Cooperative diffusion, 63
Critical field, for junction breakdown,
191, 193
Crystal structure, of Ge, Si, and GaAs
(Table), 102–103
Current crowding, in junction transistors,
229
Current gain, of junction transistors, 219
common-base, 211
common-emitter, 211
effect of collector current on, 220

Current gain, of junction transistors.
effect of reverse bias leakage
current on, 219
small-signal, 213
surface effects on, 303, 309
Current-voltage characteristics, of
gate-controlled diodes, 298–304
of junction field-effect transistors,
248–251
of p-n junctions, 172–191
of surface field-effect transistors, 321–326
Cut-off frequency, of junction
field-effect transistors, 254
of surface field-effect transistors, 328
of junction transistors, 223

Decay, of injected carriers, 119–121
Degenerate semiconductors, 101
Density, of gases, 20
of Ge, Si, GaAs, and SiO_2 (Table), 102
of states, effective, for Ge, Si, and
GaAs (Table), 102
Depletion approximation, 157, 267
Depletion region, 156
surface, 266, 293
maximum width of, 268, 293
width of, 159, 163, 166–169
Dielectric constant, for Ge, Si, GaAs,
and SiO_2 (Table), 103
Diffused junction, 3
breakdown voltage of, 196
space-charge region for, 167–169
Diffused layers, 43–58
evaluation of, 52–58
two-dimensional, 59
Diffusion, drive-in, 49–52
equation, 42
field-aided, 59–63
formulas (Table), 88
of electrons and holes, 113
of gold in silicon, 41
rate limitation, external, 65–69
solid-state, 35–83
space-charge effects on, 59–63
through silicon dioxide layers, 75-77
Diffusion current, forward bias,
183–186
reverse bias, 173, 175–177
Diffusion length, of impurities, 46
of minority carriers, 124

Index

Diffusivity, 37
  in gases, 20
  in silicon, 38–40
  in silicon dioxide, 41
  of electrons, 113
  of holes, 113
  thermal, 43
Diodes (see p-n Junctions)
Diodes, finite length (narrow base), 190
Distributed surface states or recombination centers, 302
Distribution function, Boltzmann, 99
  Fermi-Dirac, 98
Drain, junction field-effect transistors, 243
  surface field-effect transistors, 318
Drift, 106
  of ions in MOS structures, 337–340
  velocity, 106
Drive-in diffusion, 43, 49–52
Donors, 97

Early effect, 226
Effective densities, of states, 100
  for Ge, Si, and GaAs (Table), 102
Effective mass, 107
Einstein's relationship, 37, 113
Electric field, 152
  built-in, 32, 59–63, 224
Electrochemical oxidation, 22
Electron, capture, 129
  chemical potential of, 98
  concentration, 100, 104
  conduction, 93
  diffusivity of, 113
  emission, 129
  emission probability, 131
  mobility of, 108
  potential energy of, 152
  transport of, 106–114
Electrostatic potential, 152
Emission probability, of electrons, 131
  of holes, 132
Emitter, efficiency, 218
  factor, 218
  of junction transistors, 210
Emitter-dip effect, 63
Emitter-push effect, 63
Energy bands, 91

Energy gap, for Ge, Si, GaAs, and SiO_2 (Table), 102
Epitaxial diodes, breakdown voltage limitations, 199
Epitaxial growth, 7–20
  redistribution of impurities in, 78–83
Equilibrium criterion, 101
Error function, complementary, 46
  properties of (Table), 48
Etching, 8
External rate limitation, on diffusion, 65–69
Extraction of carriers, 118

Fast surface states, 283, 335–337
  in thermally oxidized silicon, 336
Fermi-Dirac distribution, 98
Fermi-Dirac distribution function, 98
Fermi-Dirac statistics, 98
Fermi level, 98
  as a function of temperature, 104
  in equilibrium, 156
  intrinsic, 99, 101
Fermi potentials, 157
Field-aided diffusion, 32, 59–63, 224
Field-effect transistors (see Junction field-effect transistors and Surface field-effect transistors)
Field-induced channel surface field-effect transistors, 330
  formulas for (Table), 333
Field-induced junction, 269, 291, 305
  breakdown voltage of, 305
Field-plate controlled p-n junction, 264
Film growth rate, 11, 17, 79
  temperature dependence of, 12
Finite length (narrow base) diodes, 190
First-order chemical reaction, 10
Flat-band voltage, 278, 281
Flux, 10, 24, 36
Forbidden gap, 91
Formulas, for field-induced channel surface field-effect transistors, 333
  for junction field-effect transistors, 259
  for junction transistors, 242
  for p-n junctions, 207
  for surface space-charge regions, 288
  in solid-state diffusion, 88
  semiconductor physics, 116

Formulas, semiconductors in non-equilibrium, 148
Forward bias, 150, 161
Forward current, diffusion, 183–186
  recombination, 186–190
  temperature dependence of, 188–189
Forward current-voltage characteristics, empirical representation, 189
  for Ge, Si, and GaAs junctions, 188–189
Fourier's law of heat conduction, 237
Four-point probe technique, 58
Frequency effects, on MOS capacitance-voltage characteristics, 274
Frequency limitation, of junction field-effect transistors, 254
  of junction transistors, 223
  of surface field-effect transistors, 328
Furnace, oxidation, 23

Gallium arsenide, intrinsic carrier concentration, 96
  properties of (Table), 102–103
Gallium arsenide p-n junctions, forward current, 189
  reverse current, 179
Gas constant, 19
Gases, properties of, 18–20
Gas-phase mass transfer, 13
  coefficient, 10, 14, 16, 18, 24
    in terms of concentrations in the solid, 25, 65, 67, 79
Gate, junction field-effect transistors, 243
Gate-controlled diode, 264, 290, 296
  current-voltage characteristics of, 298
Gate leakage current, junction field-effect transistors, 252
  surface field-effect transistors, 327
Gaussian impurity distribution, 50
Generation, of electron-hole pairs, 119
Generation current, 173, 300
Generation rate, in reverse-biased depletion region, 174, 301
  surface, 301
Germanium, intrinsic carrier concentration, 96
  properties of (Table), 102–103
  surface recombination velocity, 145
  surface-state density, 144

Germanium p-n junctions, forward current, 188
  reverse current, 178
Gettering, 201
Gold, capture cross section, 141
  diffusion in silicon, 41
  effect on resistivity, 142
Graded base region, 224
Graded channel regions, junction field-effect transistors, 253
Gradient of impurity concentration, 48, 50
Grooving, 52
Grown junction, 1
Growth rate, epitaxial film, 11, 17, 79

Henry's law, 24, 44
  constant, 25, 67
High-current effects, 227
High-level injection, 119, 227
Hole, capture, 129
  concentration, 100, 104
  diffusivity of, 113
  emission, 130
  emission probability, 132
  transport of, 106
Horizontal epitaxial reactor, 8

Ideal gas law, 19
Illumination, effect on junction reverse current, 180
Impurity, concentration gradient, 48, 50
  interstitial, 40
  scattering, 109
  substitutional, 39
Injected carriers, decay of, 119
Injection, 117
  from a boundary, 125
  high-level, 119, 227
  low-level, 119
Instabilities in MOS structures, 337
Insulator, 92
  charges in, 279–282, 337–341
  lead-silicate glass, 351
  phospho-silicate glass, 350
  silicon nitride, 352
Integrated circuits, 3
Intermediate centers, recombination-generation, 129–134

Index

Interstitial impurities, 40
Intrinsic carrier concentration, 96, 101
Intrinsic Fermi level, 99, 101
Inversion layer, 267
   mobility, 346–347
      temperature dependence of, 347
Ionic contamination of oxide, 337–340
Ionization energy, 95
Ionizing radiation, effect on fast surface states, 145
   effect on oxide space charge, 340–341

Junction field-effect transistors, 243–259
   channel, 244
   channel conductance, 245, 250
   comparison to surface field-effect transistors, 320–321
   current-voltage characteristics, 248–254
   cut-off frequency, 254
   drain, 243
   drain-current saturation, 251
   formulas for (Table), 259
   frequency limitation, 254
   gate, 243
   gate leakage current, 252
   graded channel regions, 253
   linear region, 248
   principles of operation, 244–248
   saturation, 247
   saturation region, 248
   series resistance, effect of, 256
   source, 243
   source-to-drain resistance, in saturation, 255
   transconductance, 252
   turn-off voltage, 250
Junction transistors, 208–242
   base factor, 218
   base resistance, 228–230
   base-width modulation, 226
   breakdown voltage, 230–234
   common-base breakdown voltage, 230
   common-emitter breakdown voltage, 230–234
   conductivity modulation, 227
   current components, 214
   current crowding, 229
   current gain, 219–222
   cut-off frequency, 223
   Early effect, 226

Junction transistors, emitter efficiency, 218
   emitter factor, 218
   formulas for (Table), 242
   frequency limitation, 223
   graded base region, 224
   high-current effects, 227
   maximum voltage limitations, 230–234
   minimum voltage limitations, 234–236
   one-dimensional model, 209
   planar, 208
   punch-through condition, 230
   recombination factor, 218
   saturation, 234
   surface effects, on current gain, 303, 309
   surface recombination, effect of, 218, 303
   switch, 214
   terminology, 214
   thermal limitation, 236–238
   thermal resistance, 238
   transit-time limitation, 222
   transport factor, 219

Kinetics, of oxidation, 23–31
   of recombination process, 127–134
   of vapor-phase growth, 10–13

Lattice, constant for Ge, Si, GaAs, and SiO_2 (Table), 102
   mobilities for Ge, Si, and GaAs (Table), 103
   scattering, 109
   strain, effect on diffusion, 63
Lead-silicate glass, 351
Lifetime, band-to-band recombination, 128
   effect of radiation damage on, 143
   in low-level injection, 134–136
   of excess minority carriers, 121
   within a reverse-biased depletion region, 174
Linear coefficient of thermal expansion for Ge, Si, GaAs, and SiO_2 (Table), 103
Linear region, junction field-effect transistors, 248
   surface field-effect transistors, 320
Linear oxidation law, 27

Linear oxidation rate constant, 27
  temperature dependence of, 30
Linearly graded junction, 163–166
  breakdown voltage, 195
  built-in voltage, 165
  capacitance, 171
  maximum electric field, 164
  space-charge region, 163–166
Liquid source, 43
Low-level injection, 119
  lifetime in, 134–136

Majority carrier, 98, 105
  concentration, as a function of temperature, 105
Masking, 75–77
  thickness, 77
Mass, effective, 107
Mass-transfer coefficient, 10, 14, 16, 18, 24, 67, 79
  in terms of concentrations in the solid, 25, 67, 79
Mass-transfer control, 11, 26
Maximum electric field, in linearly graded junctions, 164
  in step junctions, 158
Maximum surface recombination velocity, 140
Maximum voltage limitations, of junction transistors, 230–234
Metal, 92
Melting point, for Ge, Si, GaAs, and SiO_2 (Table), 103
Metal-insulator-semiconductor structures (see MOS structures)
Metal-insulator-semiconductor surface field-effect transistors (see Surface field-effect transistors)
Metallurgical channel surface field-effect transistors, 330
Metal-oxide-semiconductor structures (see MOS structures)
Metal-oxide-semiconductor surface field-effect transistors (see Surface field-effect transistors)
Minimum voltage limitations of junction transistors, 234–236
Minority carriers, 105
  concentrations under forward bias, 185
MIS structures (see MOS structures)

MIS transistors (see Surface field-effect transistors)
Mobility, 37
  in silicon, 109
  inversion layer, 347
    temperature dependence of, 347
  lattice, for Ge, Si, and GaAs (Table), 103
  of electrons, 108
  of holes, 108
  surface, 346
  temperature dependence of, 110
Molecular or atomic weight, Ge, Si, GaAs, and SiO_2 (Table), 102
Mole fraction, 9
MOS structures, 264
  capacitance-voltage characteristics, 271
  channel conductance, 276
  drift (instability), 337
  frequency effects, 274
  turn-on voltage, 273
MOS transistors (see Surface field-effect transistors)
Multiplication, 193
Multiplication factor, 194

Narrow-base diodes, 190
Non-equilibrium conditions, 117–148
Normally "off" surface field-effect transistors, 330
Normally "on" surface field-effect transistors, 330
np product (see pn product)
n-type conductivity, 98

One-sided step junction, 159
  breakdown voltage, 194
  capacitance, 171
  formulas (Table), 207
Orientation effect, on oxidation rate, 30
  on surface state charge, 343
Out-diffusion, 67
Oxidation, electrochemical, 22
  furnace, 23
  kinetics of, 23
  rate constants, 27
  space-charge effects on, 31
  thermal, 22
Oxide, diffusivities in, 41
  masking, 75–77

Index

Oxide, properties of, 334–355
  space charge, 337–341
  surface conduction, 347–350

Parabolic oxidation law, 27
  rate constant, 27
  temperature dependence of rate constant, 29
Phospho-silicate glass, 350
Photocurrent, junction, 180
Planar junction, 149
  breakdown voltage, 197
Planar technology, 3
Planar transistors, 208
Plasma oxidation, 22
p-n junctions, 149–207
  alloy, 2
  breakdown, 150, 191–201
  built-in voltage, 157, 165
  capacitance, 169–172
  concentration gradient at, 165
  current-voltage characteristics, 172–191
  curvature, effect on breakdown of, 197
  depth, 52
  diffused, 3
  field-induced, 269, 291, 305
  forward current, temperature dependence of, 188–189
  gate-controlled, 264, 290, 296
  grown, 1
  one-dimensional model, 150
  photocurrent, 180
  reverse current, temperature dependence of, 179
  surface effects on, 289
  transient behavior, 201–204
pn product, in equilibrium, 101
  in quasi-equilibrium, 184
  in space-charge region, 139, 184
Poisson's equation, 153
Polarization in insulators, 351–353
Potential energy of electrons, 152
Predeposition, 43–48
Properties of gases, 18–20
Properties of Ge, Si, GaAs, and SiO_2 (Table), 102–103
p-type conductivity, 98
Punch-through condition in junction transistors, 230

Q of diffused layer, 47, 50
Q of transistor base, 226
Q_{ss}, 342
Quasi-equilibrium, 184–185
Quasi-Fermi levels, 162, 184–185

Radiation damage, effect of, on fast surface states, 145
  on lifetime, 143
Radiation-induced space charge in oxides, 340
Rate constant, chemical surface reaction, 10, 25
  oxidation, 27
  temperature dependence of, 29–30
Reach-through limited junction breakdown voltage, 199
Reactor, horizontal, 8
  vertical, 8
Recombination, band-to-band, 128
  centers, 129
  continuum of, 302
  origin of, 140–145
  current, 183, 186–190
  factor, junction transistors, 218
  in surface space-charge region, 298–304
  kinetics of, 127–134
  through intermediate centers, 129–134
Rectification, 150
Redistribution of impurities, during thermal oxidation, 69–75
  in a predeposited layer, 74
  in epitaxial growth, 78–83
Resistivity, 111–113
  average, of base region, 230
  of diffused layers, 54
Reverse bias, 150, 161–163, 172–180
  leakage current, effect on current gain, 219
Reverse current, diffusion, 173, 175–180
  generation, 173–175
  of Ge, Si, and GaAs junctions, 178–179
  temperature dependence, 179
  under illumination, 180
Reynolds number, 16, 17, 23

Saturation, junction transistors, 234
Saturation current, in junction field-effect transistors, 247

Saturation current, in surface field-effect transistors, 326
Saturation region, junction field-effect transistors, 248
  surface field-effect transistors, 320
Scattering mechanisms, 109
Schmidt number, 17
Segregation coefficient, 69, 74, 77
Semiconductor, 93
  degenerate, 101
  physics, 91–116
    formulas (Table), 116
  surfaces, 263–288
    under non-equilibrium conditions, 117–148
      formulas (Table), 148
Series resistance effect, on junction field-effect transistors, 256
  on surface field-effect transistors, 329
Shockley-Read-Hall theory, 129
Silicon, diffusivities in, 38–40
  intrinsic carrier concentration, 96
  properties (Table), 102–103
  surface recombination velocity, 145
  surface state densities in, 144
Silicon dioxide, 22
  diffusivities in, 41
  masking, 75–77
  properties of, 102–103, 334–355
Silicon nitride, 352
Silicon p-n junctions, forward current, 188
  reverse current, 178
Silicon-silicon dioxide system, properties of, 334–355
Silicon tetrachloride, 7, 8
Small-signal capacitance, 169
Small-signal current gain, of junction transistors, 213
Sodium contamination, in oxides, 339
Soft breakdown, 200
Solid solubility, 44
Solid source, 43
Solid-state diffusion, 35–88
Source, junction field-effect transistors, 243
  surface field-effect transistors, 318
Source-to-drain resistance in saturation, junction field-effect transistors, 255
  surface field-effect transistors, 329

Space-charge, radiation induced, 340
  within insulator, 280
  within oxide, 337–341
Space-charge effects, on diffusion, 59–63
  on oxidation, 31
Space-charge neutrality, 101
Space-charge region, for diffused junctions, 167–169
  for linearly graded junctions, 163–166
  for step junctions, 153–163
Specific heat of Ge, Si, GaAs, and SiO$_2$ (Table), 103
Step junction, 157
  built-in voltage, 157, 161
  formulas (Table), 207
  maximum electric field, 158
  one-sided, 159
  space-charge region for, 153–163
Stagnant-film model, 13
Strong inversion, 267–268, 292
Substitutional impurities, 39
Surface conduction on oxides, 347
Surface depletion region, 266
  width, 266, 267
Surface effects, on junction breakdown voltage, 311
  on p-n junctions, 289–316
  on transistor current gain, 303, 309
Surface field-effect transistors, 264, 317–333
  channel, 318
  channel conductance, 324
  comparison to junction field-effect transistors, 320–321
  current-voltage characteristics, 321–327
  cut-off frequency, 328
  drain, 318
  drain-current saturation, 320
  effect of surface states, 327
  field-induced channel, 330
  formulas for (Table), 333
  frequency limitation, 328
  gate leakage current, 327
  linear region, 320
  metallurgical channel, 330
  normally "off", 330
  normally "on", 330
  other types, 329–331
  principles of operation, 318–321

Index

Surface field-effect transistors,
   saturation, 320
   saturation region, 320
   series resistance, effect of, 329
   source, 318
   source-to-drain resistance in saturation, 329
   surface scattering, 325
   transconductance, 326
   turn-on voltage, 324
Surface generation, current, 301
   rate, 301
Surface mobility in silicon, 346
Surface potential, 267, 293
Surface-reaction control, 11, 26
Surface-reaction rate constant, 10
Surface recombination, 121, 136–140
   centers, 301
   effect on emitter efficiency, 218
Surface recombination velocity, 124, 136, 302
   maximum, 140
   of a surface without a surface space-charge region, 139
   on Ge and Si surfaces, 145
   on thermally oxidized silicon, 145, 336
Surface scattering, effect on surface field-effect transistors, 325
Surfaces, semiconductors, 263–288
Surface space-charge region, 137, 263
   equilibrium, 264–271
   formulas (Table), 288
   non-equilibrium case, 290–296
   recombination-generation within, 298–304
Surface-state charge in thermally oxidized silicon, 342
Surface states, 144, 282, 335–337
   continuum, 302
   effect on surface field-effect transistors, 327
   in germanium, 144
   in thermally oxidized silicon, 144, 336
   single level, 336
   uniformly distributed, 336
Switch, transistor, 214

Temperature dependence, of film growth rate, 12
   of junction forward current, 188–189
Temperature dependence, of junction reverse current, 179
   of linear oxidation rate constant, 30
   of parabolic oxidation rate constant, 29
Terminology of junction transistors, 214
Thermal conductivity, for Ge, Si, GaAs, and SiO_2 (Table), 103
Thermal diffusivity, 43
   for Ge, Si, GaAs, and SiO_2 (Table), 103
Thermal limitation of transistors, 236–238
Thermally oxidized silicon, 334–355
Thermal oxidation, 22–34
   redistribution of impurities in, 69–75
Thermal resistance, 238
Thermal velocity of carriers, 108, 130
Transconductance, junction field-effect transistors, 252
   surface field-effect transistors, 326
Transient behavior of p-n junctions, 201–204
Transistors (*see* Junction transistors)
Transit time limitation, 222
Transport equation, 41
Transport factor, junction transistors, 219
Transport properties, of gases, 20
Trap, 130
Tunneling (Zener breakdown), 191
Turn-off time, for p-n junctions, 202
Turn-off voltage, junction field-effect transistors, 250
Turn-on voltage, of MOS structures, 273, 297
   of surface field-effect transistors, 324
Two-dimensional diffused layers, 59

Uniform distributions, of recombination centers, 302
   of surface states, 302
Unipolar transistors, 243

Valence band, 92
   edge, 94
Valence electrons, 91
Vapor-phase growth, 7–21
   kinetics of, 10–13
Vapor pressure for Ge, Si, GaAs, and SiO_2 (Table), 103
Velocity of carriers, drift, 106

Velocity of carriers, thermal, 108
Vertical reactor, 8
Viscosity, 14
   of gases, 20

Work function difference, effect on
   MOS structures, 278, 345

Zener breakdown (tunneling), 191